Get better grades with . . .

Scientific Notebook™

for Windows® 95 and Windows NT® 4.0
ISBN: 0-534-34864-5. **$59.95***

Breakthrough interactive software for anyone who uses mathematics!

Whether you do simple arithmetic or solve complex partial differential equations, the solutions are just a mouse-click away. This one-of-a-kind scientific word-processor—with its special built-in version of the Maple® computer algebra system—can handle arithmetic, algebra, calculus, linear algebra, and more!

The only software tool that uses correct mathematical notation!

Scientific Notebook allows you to enter equations, create tables and matrices, import graphics, and graph in 2-D and 3-D within your documents. And it's easy to send and receive documents containing live mathematics on the World Wide Web. This combination gives you a unique tool for exploring, understanding, and explaining key mathematical and scientific concepts.

*The vast majority of my students are enthusiastic about **Scientific Notebook**. They find the difference between using **Scientific Notebook** and using a graphing calculator something like the difference between riding in a donkey cart and taking a ride in the space shuttle.* —Johnathan Lewin, Kennesaw State University

To order a copy of *Scientific Notebook*, please contact your college store or place your order online at: http://www.scinotebook.com or fill out the order form and return with your payment.

* Call after 12/1/97 for current prices: 1-800-487-3575

ORDER FORM

______Yes! Send me a copy of ***Scientific Notebook™ for Windows® 95 and Windows NT® 4.0*** (ISBN: 0-534-34864-5)

_____Copies x $59.95* =___________

Residents of: AL, AZ, CA, CT, CO, FL, GA, IL, IN, KS, KY, LA, MA, MD, MI, MN, MO, NC, NJ, NY, OH, PA, RI, SC, TN, TX, UT, VA, WA, WI must add appropriate state sales tax.

Subtotal ___________
Tax ___________
Handling $4.00
Total Due ___________

Payment Options

_______ Check or money order enclosed

Bill my ____VISA ____MasterCard ____American Express

Card Number: ______________________________________

Expiration Date: ______________________________________

Signature: ______________________________________

Please ship my order to: *(Credit card billing and shipping addresses must be the same)*

Name ______________________________________

Institution ______________________________________

____

Street Address______________________________________

____

City ______________________ State __________ Zip+4______________

Telephone ( )______________________ e-mail ______________________

Your credit card will not be billed until your order is shipped. Prices subject to change without notice. We will refund payment for unshipped out-of-stock titles after 120 days and for not-yet-published titles after 180 days unless an earlier date is requested in writing from you.

Mail to:

Brooks/Cole Publishing Company
Source Code 8BCTC011
511 Forest Lodge Road
Pacific Grove, California 93950-5098
Phone: (408) 373-0728; Fax: (408) 375-6414
e-mail: info@brookscole.com

* Call after 12/1/97 for current prices: 1-800-487-3575

10/98

Precalculus in Context

Projects for the Real World

Precalculus in Context

Projects for the Real World

SECOND EDITION

Marsha J. Davis
Eastern Connecticut State University
Judith Flagg Moran
Trinity College
Mary E. Murphy
Smith College

Brooks/Cole Publishing Company
I(T)P® An International Thomson Publishing Company

Pacific Grove • Albany • Belmont • Bonn • Boston • Cincinnati • Detroit • Johannesburg • London
Madrid • Melbourne • Mexico City • New York • Paris • Singapore • Tokyo • Toronto • Washington

Sponsoring Editors: *Margot Hanis and Beth Wilbur*
Editorial Associate: *Nancy Conti*
Marketing Team: *Laura Caldwell and Christine Davis*
Art Editor: *Lisa Torri*
Production Editor: *Mary Vezilich*
Permissions Editor: *May Clark*
Cover Design: *Lisa Henry*
Cover Photo: *Yoshinori Watabe/Photonica*
Photo Research: *Kathleen Olson*
Typesetting: *Integre Technical Publishing Co., Inc.*
Printing and Binding: *Patterson Printing*

For more information, contact:

BROOKS/COLE PUBLISHING COMPANY
511 Forest Lodge Road
Pacific Grove, CA 93950
USA

International Thomson Publishing Europe
Berkshire House 168–173
High Holborn
London WC1V 7AA
England

Thomas Nelson Australia
102 Dodds Street
South Melbourne, 3205
Victoria, Australia

Nelson Canada
1120 Birchmount Road
Scarborough, Ontario
Canada M1K 5G4

International Thomson Editores
Seneca 53
Col. Polanco
11560 México, D.F., México

International Thomson Publishing GmbH
Königswinterer Strasse 418
53227 Bonn
Germany

International Thomson Publishing Asia
221 Henderson Road
#05–10 Henderson Building
Singapore 0315

International Thomson Publishing Japan
Hirakawacho Kyowa Building, 3F
2-2-1 Hirakawacho
Chiyoda-ku, Tokyo 102
Japan

Printed in the United States of America

10 9 8 7 6 5 4 3 2 1

Library of Congress Cataloging in Publication Data
Davis, Marsha Jane, [date]
Precalculus in context : projects for the real world / Martha J. Davis, Judy Flagg Moran, Mary E. Murphy. — 2nd ed.
p. cm.
ISBN 0-534-35232-4 (paper)
1. Functions. I. Moran, Judy Flagg. II. Murphy, Mary E. III. Title.
QA331.3.D38 1998
512.9′6′078—dc21 97-43190
CIP

Cover: Time-lapse photograph of star trails resulting from the earth's rotation about its axis.

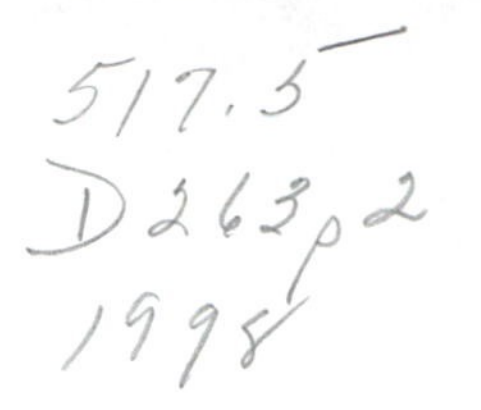

Preface

Precalculus in Context: Projects for the Real World serves as a valuable companion to any precalculus text. The labs, projects, and explorations are designed to form an integral part of a college or high-school precalculus course. Except for two sections on multivariable functions, the topics are those you would expect to find in a one-semester college course and are presented in the traditional order, beginning with linear and quadratic functions and ending with exponential, logarithmic, and trigonometric functions.

Here, however, the similarity ends. We have attempted to avoid the compartmentalized structure of traditional precalculus by weaving several themes through the book:

- mathematical modeling of real-life phenomena
- choice of a convenient and meaningful scale (including exponential and logarithmic scales)
- constant and nonconstant rates of change (and how to interpret them in context)
- relationships between algebraic statements and geometric representations

And we have based our pedagogy on four principles:

- writing about mathematics deepens understanding
- student exploration is at least as valuable as teacher explanation
- graphing technology widens the range of questions we can consider
- collaboration, rather than competition, promotes genuine learning

The second edition contains two new labs and many new explorations and projects. In response to our changing world, to advances in graphing technology, and to the experiences of our students, we have revised and updated everything that we kept from the first edition.

Vive la Difference

The spirit of the book is one that may be new to a student trained to think that doing mathematics means manipulating an algebraic formula, finding an answer, and checking it in the back of the book. This book *has* no "back"! There are, however, results that are correct and others that are not; there are conclusions and explanations that are better than others. There might be several reasonable ways to answer a question. One-sentence or one-number answers are not appropriate for the activities in this book. The explorations encourage students to use a graphing utility to investigate mathematical relationships and discover patterns. Most of the labs and projects plunge the students into a real-life situation and ask them to analyze the mathematics they find there. They should always be checking among themselves (rather than in the back of the book) to be sure that their results jibe with reality. Although some students may resist this approach, many others will find themselves (much to their own surprise) blossoming in a mathematics class for the first time in their lives.

What's more, the manner in which this book is intended to be used is strikingly different from that of a traditional precalculus text or workbook. Unlike other lab manuals, this is not a step-by-step instruction book. Each of the thirteen labs is designed for group collaboration, and each lab culminates in a written report submitted by the group. The book guides students in producing a report in which they explain in understandable prose the mathematical ideas they've been investigating, relating them to the context of the lab. The lab report is an opportunity for students to demonstrate both their grasp of the concepts and their creativity in making those concepts accessible, not simply to their mathematics instructor, but to any educated reader.

You will observe that there are no numbers on the questions posed within each lab. This was deliberate. We have found that, regardless of what we say, if there are numbered questions within a project, some students will think that the *real* goal is to find the right answers to all the numbered questions. We want, instead, to give them a format that encourages them to use the questions as guides in proceeding through the investigation and as hints for what they might want to include in their lab report. Their report should be a unified discussion, not a series of answers.

How the Book Is Organized

This manual is divided into eleven sections, each corresponding to a topic from precalculus. Each section begins with a **lab** (or two labs), and is followed by several associated activities that extend the section topic. In nearly every section, the first of these activities is a guided **exploration** of a particular type of function. Some sections contain an additional exploration that allows students to investigate a particular feature of a function, such as its long-term behavior. Grouped with the explorations are the **projects**: activities that can serve as homework assignments, in-class group projects, take-home test questions, or extra-credit problems. Most of the projects are context-based; several are direct follow-ups of a particular lab, and are labeled as such.

For the labs, we expect students to work in groups under the guidance of an instructor or a teaching assistant. For the projects and explorations, students can work either collaboratively or independently, according to the wishes of the instructor.

Use with Any Graphing Utility

Students need ready access to a graphing utility—either a graphing calculator or a computer graphing program. Throughout the book, we will use the term **grapher** to refer to the graphing utility. We have included tutorials for the TI-81/82, TI-85/86, and TI-92/93 graphing calculators, but we do not mean to imply that these particular calculators are more appropriate for the material than other graphers.

The Instructor's Guide

For many instructors, this approach to precalculus constitutes a significant departure from traditional methods. The *Instructor's Guide*, available to all adopters of the lab manual, is full of hints for using each lab, project, and exploration, suggestions for running a precalculus laboratory, and a lab-by-lab outline of features to look for when grading lab reports. We don't want to prescribe rigid requirements, but we do want to give first-time users the benefit of our experience in teaching with these materials. Admittedly, implementing this laboratory approach to precalculus requires extra effort from both instructors and students, but the rewards—excitement in the classroom, active student involvement, and opportunities to be creative—are well worth the energy expended.

Have fun!

ACKNOWLEDGMENTS

Our deepest thanks go to our students, whose questions, difficulties, interests, and suggestions inspired us to write this second edition of *Precalculus in Context*, as well as to our colleagues Jack Lutts (University of Massachusetts, Boston), Sandra Rhoades (Keene State College), and Helen Doerr (Syracuse University), who have used these materials in their teaching and generously shared their ideas. We continue to benefit from the research and editorial assistance of Anne Kaufman; her dedication made the first edition possible and the second edition manageable. Jim Douglas helped with titles and a myriad of other details. Lee Meadows Jantz furnished the data from the Forensic Anthropology Data Bank at the University of Tennessee. We are grateful to our team at Brooks/Cole—Margot Hanis, Beth Wilbur, Mary Vezilich, Kelly Shoemaker, and May Clark—for understanding the philosophy behind the book and allowing us the freedom we needed to write an unconventional text. Finally, we thank the New England Consortium for Undergraduate Science Education (NECUSE) for helping to fund our initial efforts to transform precalculus into a laboratory course.

For the Student

Dear Mathematician,

Yes, we mean you!

We are inviting you to experience mathematics in a way that might be new to you, but that is very familiar to a mathematician or a scientist.

This manual presents a series of situations for you to investigate with your lab partners. We expect and encourage you to be true researchers: to use experimentation, discovery, and common sense seasoned with inspired guessing to analyze the situation presented in each lab.

Like mathematicians and scientists, you will find working with your colleagues invaluable. The members of your lab group will bring distinct viewpoints and talents to the week's problem. We hope you will find that talking about mathematics with one another and then writing up your group's ideas in a lab report will clarify and refine your own understanding. In the "real world," the world we want you to investigate in these labs, mathematics is seldom a solo act.

Also, as in the real world, there might be no single correct answer to certain questions posed in a lab. Some conclusions may be formulated more clearly than others, some approximations may be more precise, but a lab report isn't "wrong" or "right" any more than is a paper you write for an economics course. Like your economics teacher, your mathematics instructor will be interested in how well you analyze data, how adept you are at formulating conclusions, and how coherently you express your results.

This may not feel like math to you. (As one of our former students protested, "This isn't math! You have to *think* about it!") We hope that at the end of this course you will appreciate that, in fact, mathematics *does* require that you think—that you reason well and thoughtfully—but also that you wonder. One of the best ways for a mathematician or any researcher to begin a sentence is, *"What if . . . ?"*

Your colleagues,

Marsha Davis

Judith Moran

Mary Murphy

Writing a Group Lab Report

The weekly lab report is one of the most important components of your precalculus course. It is an excellent means of gauging how well you and the other members of your group understand the material and are able to use the concepts you're learning. Each report represents the joint work and conclusions of the group and is typed by the *scribe.* The office of scribe rotates among the group members so that each student ends up typing approximately the same number of lab reports during the semester.

As soon as the first draft of the report is written, the scribe makes a copy for each member of the group. The others review the draft, making corrections and suggestions, and perhaps consulting the instructor about points that need clarification. The scribe then revises the report to reflect the consensus of the group. Be sure that any lab report that you submit has the approval of the group and that you have adequately reviewed anything that another student submits in your name.

You should also receive and keep a copy of the final version of your group's report with the instructor's comments and evaluation. The finished, graded, reports will form an important body of notes to which you will refer during the semester and from which you will want to study at exam time.

Each lab group will design its own variation on the above procedure. The important thing is that responsibility be shared and that every member be fully apprised of what the group is doing.

On the last page of each lab you will find instructions specific to that particular investigation. Below are some general guidelines that apply to every lab.

A lab report is not a list of answers to the questions posed by the lab sheets. It is an *essay* communicating mathematical ideas. **Anyone should be able to pick up your report and understand what the lab was about.** Think of your audience as someone familiar with precalculus mathematics who has not read the lab manual. You need to write in complete sentences, explain enough so the reader will understand, and structure the paragraphs so that the reader can follow the discussion. This does not mean that the report needs to be very long; it does mean that it must be able to stand on its own.

A good lab report contains the following:

- an introduction for your reader
- the mathematical concepts involved
- an outline of your procedure
- anything you learned along the way
- summary and conclusions

Graphs can do much to clarify what you're saying. They needn't be elaborate, but the axes should always be labeled and the units, if appropriate, should be specified. Graphs can be woven into the text or gathered at the end of the report.

At first, you might find yourself highlighting the questions in the lab and then going about the business of answering those questions. Fine. But that, in itself, doesn't make a lab report. Imagine yourself a tailor: you cut out the pieces of fabric and put them together with large basting stitches; then you sew the seams and remove the bastings. Your final suit shouldn't have basting stitches in it. Some of the questions in the lab sheets are like basting stitches: answering them helps to shape your work and keep you from straying too far off the track. The finished product doesn't need even to mention those questions; the fact that you produced an elegant piece of work shows that you put things together correctly.

Other questions, however, do need to be answered in the lab report. How can you tell which questions are basting stitches and which are essential construction details? Experience will help. So will the specific guidelines at the end of each project. If a lab asks for a specific calculation ("What is the volume when L = 44 in.?"), that's probably just an illustration for your benefit, and the answer ("V = 7744 cu. in.") doesn't need to be in the report unless you're using it to illustrate a concept. If, however, the lab asks a big question ("Why does one shape yield more volume than the other?"), your answer will form an essential part of the report and should be incorporated in your discussion.

The lab pages help to lead you, by means of questions, from specific examples to general conclusions. Your lab report, in contrast, might do better to begin with the conclusions and then back them up with algebraic evidence and illustrate them with examples and graphs. In other words, though you conduct the investigation in the order suggested by the lab pages, you might write about it more effectively by switching things around.

There is room for plenty of creativity, because you get to decide the most effective way of presenting your ideas. Take the opportunity to read other groups' reports; you'll notice that no two are the same (nor should they be).

Take your lab work seriously, but don't be solemn about it. Creativity and even humor have a place in a lab report. Many precalculus students find that they learn more mathematics from the labs than from any other activity during the semester. Writing about mathematical ideas that you truly understand can be an enjoyable and satisfying experience. We hope that it becomes yours.

Contents

Lab 1A: Fahrenheit

Some Like It Hot

PREPARATION

After counting, one of the most basic uses of numbers is measurement. English-speaking countries have inherited a particularly arbitrary system of measurement. One example is the use of degrees Fahrenheit to measure temperature. The Fahrenheit scale was established in the early eighteenth century by Daniel Gabriel Fahrenheit, a German physicist. Fahrenheit obtained a reading for the uppermost fixed point on his scale by placing a thermometer under the armpit of a healthy man; he obtained the lowest, 32, by using an ice and water mixture to determine the freezing point of water. The number 0 approximated the temperature of an ice and salt mixture (which was widely considered to be the coldest possible temperature), so all readings on a Fahrenheit thermometer were assumed to be positive. Fahrenheit arbitrarily assigned to the upper and lower fixed points the values 96 and 32 to eliminate "inconvenient and awkward fractions."

By contrast, the Celsius scale, like our number system, is based on the number 10. The fixed points, corresponding to the freezing and boiling temperatures of water, are labeled 0 and 100, respectively. In United States, the campaign to "convert to metric" has been waged for decades. One problem is that, in order to use a system of measurement effectively, a person needs to internalize the scale. We know that a room temperature of 70 degrees Fahrenheit feels comfortable, but that 60 degrees requires a sweater and 50 degrees a coat. But if the weather forecaster predicts a high of 17 degrees Celsius, many of us would need to convert this temperature into our internalized Fahrenheit system in order to anticipate the weather correctly.

Many banks and businesses provide time and temperature signs with temperature given in both Fahrenheit and Celsius degrees to help us internalize the Celsius scale. Suppose that, over the course of the semester, you collected the following data from an electronic sign on a bank in your town. (Note that the sign reports only integer values for the temperatures!)

Degrees Celsius	Degrees Fahrenheit
−20	−4
−4	25
0	32
5	41
12	54
13	55
24	75
34	93

One way to understand the relationship between the two scales is to draw a graph. On graph paper, draw a horizontal axis representing degrees Celsius and a vertical axis for degrees Fahrenheit, and plot the points corresponding to the numbers in the table. Do the points make a pattern? If so, describe it.

We call the data in the table *discrete:* each value given by the sign is separated by at least one degree from any other. However, we know that temperature varies *continuously:* for the temperature to rise from 3 degrees to 4 degrees, it must pass through all intermediate real values (even $\sqrt{11}$ and π!). Thus we are justified in approximating our points by a continuous curve. Do this; then use your graph to estimate what Celsius temperature corresponds to 0 degrees Fahrenheit.

Draw two more graphs of the same data, as follows. First, change the scale of the horizontal axis by doubling the distance you used on the original graph to represent one unit. (Leave the vertical axis unchanged.) Second, halve the distance used to represent one unit on the original graph. How do your three graphs differ from one another? Which features remain unchanged?

Bring your graphs with you to your first lab. In Lab 1A, you will be investigating the relationship between temperature measured in degrees Celsius and temperature measured in degrees Fahrenheit. You will also use your grapher (that is, your graphing calculator or a computer program) to further explore the effects of altering the scale on one of the axes of your graph.

"LET'S GO OVER TO CELSIUS'S PLACE. I HEAR IT'S ONLY 36° OVER THERE."

THE FAHRENHEIT LAB

In your preparation for this lab, you drew three graphs of the relationship between temperature measured in degrees Celsius and temperature measured in degrees Fahrenheit. Compare your graphs with those of your lab partners to make sure you agree on how to represent the data given in the table in the preparation section. Do you all agree on what changes and what remains the same when one of the scales is altered?

In the table, both degrees Celsius and degrees Fahrenheit vary, but the variation is not random. In fact, if we know the temperature in degrees Celsius, we can find the Fahrenheit temperature. Because the temperature measured in degrees Fahrenheit is uniquely determined once we know the Celsius temperature, we can say that degrees Fahrenheit is a **function** of degrees Celsius. Let the letter C represent the number of degrees Celsius and F the number of degrees Fahrenheit. We call F and C **variables** because their values change. We would like to write the relationship between F and C explicitly using algebra. For F to be a function of C, this relationship would take the form

$$F = \text{some algebraic expression involving } C$$

An equation written in this form will show the way in which the value of F depends on the value of C, so we call F the **dependent variable** and C the **independent variable**. We want to find an equation of this form to serve as a **model** for the data in our table. (The equation is an idealization of the data; the numbers in the table might not fit the equation exactly.) That is, we want to perform the same algebraic operation on each number in the first column and obtain (approximately at least) the corresponding number in the second column.

A very simple relationship between F and C would be one in which, given a value for C, we obtain the corresponding value for F by multiplying by a fixed constant. If k represents this constant, then the equation we're looking for would be of the form $F = k \cdot C$. If the relationship between F and C is of this form, we say that F is **proportional** to C.

Here's an everyday example: the amount of sales tax is proportional to the price of an item. If a state imposes a sales tax of 7% and the item costs \$22.65, the amount of sales tax is .07 times \$22.65, or \$1.59. In general, if the item costs D dollars, the amount of the sales tax is $.07D$.

Do you think F is proportional to C? Using values from the table, along with our mathematical definition of *proportional*, construct an argument to support your answer.

Now, try to construct an equation that models the relationship between F and C. We are deliberately not giving you an exact recipe for doing this. Talk with each other about how you might use the values in the table to write an algebraic relationship between the two variables. A worthwhile investigation often involves making educated guesses, trying ideas that don't work, improving those ideas, and testing your conjectures. (Hint: the third row of your table is very helpful.)

When you agree on an equation, test it with all the C-values in the table. Are you satisfied with the F-values it produces? Remember that not all will exactly match the ones in the table. Do you understand why some values are not exact?

What does your model predict for the value of C corresponding to $F = 0$? In the preparation section, you used your graph to estimate that C-value. Does the algebraic answer agree with your graphical estimate? Explain any difference between them.

The **domain** of a function is the set of allowable values for the independent variable. We will sometimes call the independent variable the *input* variable because, if we were using a calculator to compute F, we would input -20 for C, for example, to obtain the value -4 for F.

The domain of the function defined by the table consists of eight numbers. What are they?

The domain of the function defined by the *equation* that models the data is not the same. The equation can be used to compute F given *any* value of C, so the domain of that function is $\mathbb{R}$, the set of all real numbers.

Use your grapher to graph the algebraic relationship between F and C that you found. Adjust the viewing window so that the graph looks like the first one you drew in preparation for this lab. Then answer the following questions. You will probably need to use a variety of viewing windows to observe different parts of the graph.

Fahrenheit believed that 0° F was the coldest possible temperature. (Those of us living in New England know better!)
What value for C gives an F-value of approximately -10?
As C increases, what happens to F?
As C increases, does F change more slowly than C or more rapidly?
As C decreases, what happens to F?
Is F always larger than C? (Are you sure?)
Is there any temperature for which $F = C$? If so, what is this common value?
How did your grapher help you to answer these questions?
When $C = 0$, $F = 32$. When $C = 5$, $F = 41$. From the graph, estimate F when $C = 10$.
How much did F change from the value it had when C was 5?
How much do you think F will change between $C = 10$ and $C = 15$? Does your graph confirm this answer?
Complete the following: for this function, a change of ______ in the value of C produces a change of ______ in the value of F.
If the Celsius temperature goes up one degree, what happens to the Fahrenheit temperature?

Now look at the algebraic relationship between F and C that you postulated. Do you see a way of "reading" the information about relative changes in F and C from the equation?

From your model, you see that some integer values of C produce fractional values for F and vice versa. The electronic sign, however, displays only integer values for temperature. There are two conventions for doing this: rounding off, which replaces a rational number with the closest integer, and truncation, which drops the fractional part of a number. Which method do you think the electronic sign uses? Use specific entries from the table to justify your answer.

THE LAB REPORT

Your report should be an essay discussing your work in Lab 1. It should be accessible to the general reader, not just to the instructor and other members of the class.

- Give the algebraic relationship you determined to model the relationship between F and C. Show how you found this relationship.
- Discuss whether or not F is proportional to C, using data from the table to justify your answer.
- Include a careful sketch of the graph of the function you determined. Then present a second graph of the same function with a different scale for the horizontal axis. Explain which features of the graph are altered by a change of scale and which features remain the same.
- Explain how you can "read" the algebraic relationship between F and C from the graph. (Alternatively, explain how the constants that appear in the algebraic formula affect the graph of the function.)
- This lab presents two related functions—the one defined by the table of values and the one defined by the algebraic formula—that differ in several ways. What differences do you see between the two functions?
- Which method, rounding off or truncation, does the electronic bank sign use? Show numerical evidence.

The preceding list is meant as a guide, not a straitjacket. Don't omit any of the points, but feel free to write about them in whatever order seems logical to you, and to include anything else that you might have found interesting or significant while you were doing the lab.

Lab 1B: Jurassic

Timing is everything

PREPARATION

Cecil A. DeMille has a dilemma: his roommates have given him the responsibility of taping *Jurassic Park,* which is to be shown on TV tonight. Cecil has to leave for his precalculus class shortly after the movie starts, so he won't be around to pause the machine during commercials. The entire show, then, will run for three hours, and he needs to get the whole thing on tape. That means programming the VCR before he leaves the house.

Video recording cassettes usually provide a choice of three recording speeds. At the highest speed (SP), recording quality is at its highest, but the tape will hold only two hours of video. At the intermediate speed (LP), the tape will hold twice as much, but with some loss of quality. At the lowest speed (EP), a full six hours of material will fit on a single cassette, but the quality of the reproduction is less good than in either of the other modes.

Recording mode	Tape capacity
SP	2 hours of programming
LP	4 hours of programming
EP	6 hours of programming

So here's Cecil with a three-hour movie, and he must make a choice of speeds. With SP, the whole movie won't fit on one cassette. He could record the entire movie using LP or EP, but the quality would not be the highest possible and there would be wasted tape. Thinking about how to utilize that extra tape, he has an idea: why not start out at SP, assuring the highest quality, and switch over to EP just in time to squeeze the last part of the movie onto the cassette?

The question is *when*?

What is the maximum amount of time that it is safe to record at SP without risking the loss of part of the movie? Cecil's roommates are counting on him not to lose the final 10 minutes of *Jurassic Park.*

Cecil himself is making frantic calls to electronics stores, but they're already closed. He wishes he could ask his precalculus instructor, but it's too late now.

But wait! Maybe he can use some of the math he's been learning! Just in time, he comes up with a solution. He will make the best possible recording, not lose a minute of *Jurassic Park,* and still get to class on time.

To get an idea of the time at which Cecil will switch speeds, complete the following tables.

SP Mode

Recording time (hrs)	Fraction of movie recorded	Fraction of tape used
0	0	0
0.5		
1.0		
1.5		
2.0	2/3	1

EP Mode

Recording time (hrs)	Fraction of movie recorded	Fraction of tape used
0	0	0
0.5		
1.0		
1.5		
2.0	2/3	1/3
2.5		
3.0	1	1/2

Now use the information in your tables to decide how long Cecil should record in SP mode before switching to EP mode. After how many hours should he switch modes?

In this case, it is possible to determine the answer directly from the entries in your tables. That will not always be so, because not every show is three (or some other "nice" number) hours long. During the lab, you will learn an algebraic process for figuring out the optimal switching time, and you will develop some general methods to apply in other situations.

THE JURASSIC LAB

In preparation for this lab, each of you filled in two tables and decided upon the length of time Cecil could allow the VCR to run at SP before switching over to EP in order to accommodate the entire length of a three-hour movie. Recall that the SP mode yields two hours of recording time on a cassette, while the EP mode gives six. Spend a few minutes comparing your answers and talking about your reasoning processes. Did anyone attempt an algebraic solution?

This is really a problem about *rates.* The VCR has three different rates at which it records, and each specific rate determines the quantity of material that can be recorded on a single cassette.

The Three-Hour Movie

Let's take LP, the intermediate speed, as an example. A tape recorded in LP mode holds four hours of programming. Each hour of programming, therefore, occupies one-fourth of the tape. That is, the tape fills at the *rate* of $\frac{1}{4}$ *tape per hour.*

If Cecil runs the VCR in LP mode for two hours, what fraction of the tape is full?

If he runs it for three hours, what fraction of the tape is full?

At what rate does the tape fill in SP mode?

If he runs the VCR in SP mode for two hours, what fraction of the tape is full?

If he runs it for half an hour, what fraction of the tape is full?

At what rate does the tape fill in EP mode?

If he runs the VCR in EP mode for an hour and a half, what fraction of the tape is full?

Now let's set this up algebraically: Let T represent the switchover time; that is, T is the number of hours during which he lets the machine run in SP mode. The total length of the program is three hours. What algebraic expression represents the time remaining for the machine to run in EP mode?

Above, you computed rates (*tape per hour*) for the two modes he's using: one fraction for SP and a different fraction for EP. Write an equation expressing the fact that

$$\text{SP rate} \cdot \text{SP time} + \text{EP rate} \cdot \text{EP time} = \text{one whole tape}$$

(Pay attention to units; notice how *hour* cancels out and leaves only *tape.*)

Solve that equation. Does your answer agree with the one you obtained from your tables? If so, you know a way of recording a three-hour movie on a standard cassette. Big deal! But one of the great virtues of mathematics is that it allows us to set up models and make generalizations that apply to other situations. So, let's now find out how we might handle a movie of a different length.

The x-Hour Movie

The previous section should give you an idea for generalizing this procedure. We'll still let T represent the number of hours to run the tape in SP mode, and we'll let x be the length, in hours, of the program we want to record. T, our switchover time, depends upon x; we can think of T as a *function* of x, because the value of x will determine the value of T.

Look back at the original equation you wrote for the three-hour movie. All you need to do is to replace "3" with x. You still have an equation, but it contains two variables instead of one, so you cannot solve it for a numerical answer. Solve it instead for T.

In its new form, $T = \ldots$, your equation now defines T as a function of x. Whenever a numerical value is subsituted for x, a value of T is uniquely determined. In fact, mathematicians would now write $T(x)$ rather than just plain T, emphasizing that T depends upon x in the particular way specified by the expression following the equals sign.

Check your $T(x)$ by evaluating it for $x = 3$ hours (the length of Cecil's movie). It should give the same answer that you found when you solved the three-hour problem.

Now try it for $x = 2$ hours. You ought to obtain $T = 2$. (What does $T = 2$ mean in this context, and why does it make sense?)

If you obtained both of the above answers from your $T(x)$ function, you have probably written the model correctly.

Sketch a graph of $T(x)$. You should see a line sloping downward from left to right. Interpret each of the following facts in the context of making a recording. (Your interpretation should go something like this: "If the movie is ____ hours long, we need to switch from SP to EP after ____ hours.")

- The graph passes through the point $(3, 1.5)$.
- The graph passes through the point $(4, 1)$.
- The graph passes through the point $(5, 0.5)$.

Here are some additional questions to consider:

The Godfather is showing on public TV (no commercials!); its running time is 171 minutes. When should you switch modes? (Be careful of units.)

The graph passes through the point $(1, 2.5)$. What, if anything, does that fact mean for our model?

What is the x-intercept of the graph? Does it have significance for the model? If so, what?

The graph passes through $(2, 2)$. What does that mean in this context?

Where does the graph intercept the vertical axis? Does it have any meaning for our model? Explain.

When you answered some of the previous questions, you probably realized that not every point on the graph has meaning in our model. As an abstract function, $T(x)$ has as its domain $\mathbb{R}$, the set of all real numbers, because any real-number value can be used as input. As a mathematical model for the VCR, though, $T(x)$ has a limited domain. What is the domain of the model? (That is, what set of x-values result in a set of $T(x)$-values that make sense in this context?) Is the domain of the model a discrete set (that is, is each x-value in the domain separated from the next one by a measurable amount?) or a continuous set?

Corresponding to the domain of $T(x)$ is its range. In the case of the abstract function, the range is $\mathbb{R}$, because there is no real number that cannot result as output. In the case of the model, the range is limited. Given the domain you decided upon, what is the range of the model? (The model's range must correspond to its domain.) Is the range discrete or continuous?

On your sketch of the graph, highlight the portion that represents the model. On each axis, write a descriptive label: *length of movie (hours)* on the x-axis and *recording time at SP (hours)* on the T-axis.

What is the slope of the line? The slope of a graph also has units: vertical units per horizontal unit. What are the units for the slope of $T(x)$? What do they mean for this model? (It might help to begin your sentence, "For every increase of one hour in the length of the movie, . . .")

Is T proportional to x? How do you know? (Recall that two variables are proportional if there exists a constant multiplier that converts the first into the second.)

Are *changes* in T proportional to *changes* in x? That is, does a change of one unit in x always produce a fixed amount of change in T? Your answers to the questions about slope should help you with this question.

Other Schemes

Combining the SP and the EP modes are not the only recording possibilities. Investigate the effects of using SP and LP in sequence. By adapting the procedures and the reasoning that you already used, write and graph a second mathematical model, $U(x)$, which combines SP and LP modes. Determine its domain and range. Give its slope and explain what it means in context.

Now combine LP and EP to produce a third mathematical model, $V(x)$. Determine its domain and range; discuss the meaning of its slope.

We have not exhausted the possibilities, but we probably have exhausted *you.* Cecil could have decided to use all three speeds in sequence. But the mathematics gets more complicated because we need an extra variable, so we will now lower the curtain on our show and raise the house lights.

THE LAB REPORT

Suppose that you're the manager of an electronics store. Customers keep asking you and your employees for advice on how to best record movies and other long TV programs. You don't want to figure out everything from scratch each time somebody asks, and you don't want your clerks giving incorrect advice. Write a brief guide to which your clerks can refer when faced with such questions.

As an appendix to the guide, provide the mathematical background:

- Draw a graph of each of the three functions you devised.
- Highlight the portion of each graph that represents a mathematical model for a VCR taping.
- Give the domain and the range of each model, explaining what they mean in this context.
- Give the slope of each model and explain what it means in terms of making a recording.

Even though you will be discussing mathematical terms such as domain, range, and slope, try to use as much everyday language as possible so that your clerks would have a prayer of understanding what this has to do with taping movies.

Section 1—Projects and Explorations

1.1: HIT THE SLOPES!—Exploring Linear Functions

The purpose of this activity is to help you familiarize yourself with the computer or calculator grapher that you'll be using during this course. If you're already at ease with your grapher, you might investigate some more complicated functions as well as the ones suggested here.

Adjust your viewing window so that the x- and y-axes have the same scale—that is, the distance between 0 and 1 on the x-axis is the same as the distance between 0 and 1 on the y-axis. On many graphers, this adjustment is called **square scaling**.

1. Overlay the graphs of the functions $f(x) = x$, $g(x) = x + 1$, and $h(x) = x + 2$. Describe in words the pattern that you see. Sketch the graphs of these three functions on a single set of axes.

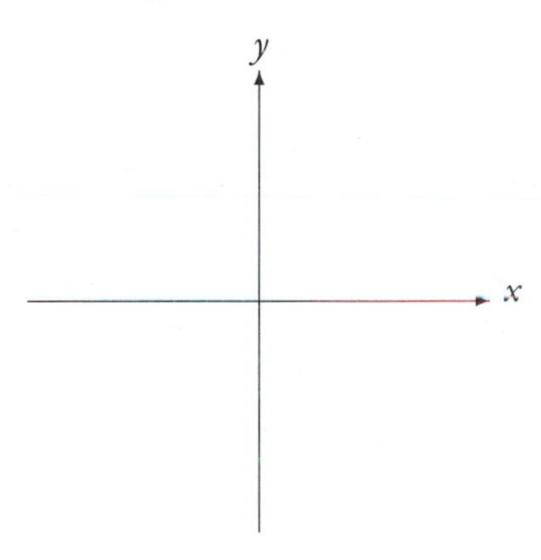

2. Overlay the graphs of the functions $f(x) = x$, $g(x) = 2x$, and $h(x) = 4x$. Sketch the graphs of these three functions on a single set of axes. Describe in words the pattern that you see. Which of the graphs rises most steeply? Which rises least steeply?

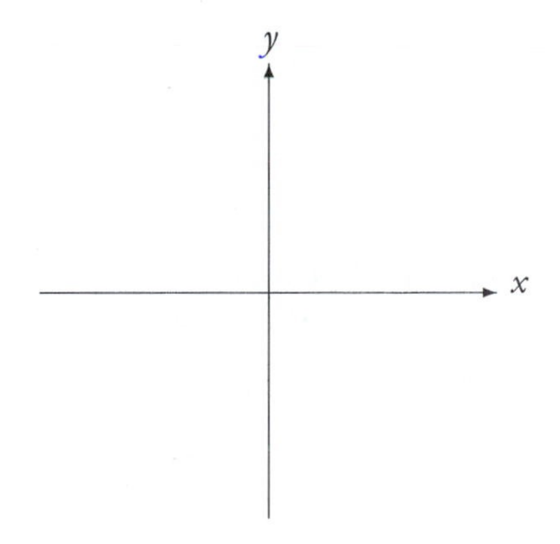

3. Overlay the graphs of the functions $f(x) = x+1$, $g(x) = 2x+1$, and $h(x) = 4x+1$. How does this set of graphs compare to the set in the previous problem?

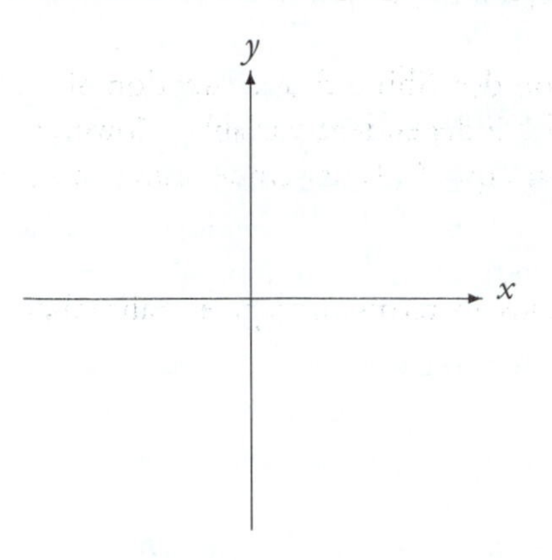

4. Overlay the graphs of $f(x) = x$ and $g(x) = -x$. Sketch what you see. Do these lines appear to be perpendicular? (Perpendicular lines meet at a ninety-degree angle.)

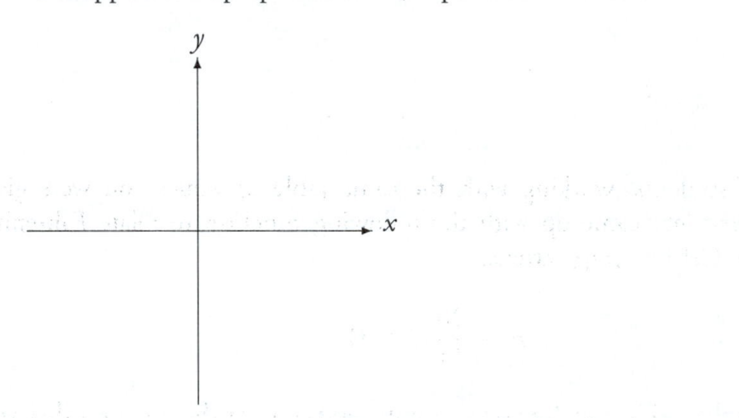

5. Graph $f(x) = -2x$. Would a line perpendicular to $f(x)$ have positive or negative slope? Overlay the graph of $g(x) = 2x$. (Be sure that the x- and y-axes still have the same scales.) Would a line perpendicular to the graph of $f(x)$ have larger or smaller slope than the line $y = 2x$?

6. Experiment with your grapher (a trial-and-error approach is fine) and find a line that appears to be perpendicular to the graph of $f(x) = -2x$. Write down the equation of this line.

1.2: HAVE IT YOUR WAY . . .

This activity is a follow-up to Lab 1A

1. In the Fahrenheit lab, you wrote an equation describing F as a function of C; that is, F was the dependent variable and C the independent variable. However, each Fahrenheit temperature also determines a unique Celsius temperature, so that we can also write C as a function of F.

   Write the equation that gives degrees Celsius in terms of degrees Fahrenheit and show how you determined it.

2. One group of students, working with the same table of values you were given in the Fahrenheit lab, came up with the following function to relate Fahrenheit temperatures to Celsius temperatures.

$$F = \frac{20}{11}C + 32$$

   Using your grapher, view this function simultaneously with the function that your group determined.

   By substituting all the values from the table given in the preparation, determine how well this second function models the data. Do your results suggest any conclusions about models of real-life situations?

1.3: A BIG MOOSETAKE

The moose populations in Vermont, New Hampshire, and Maine increased sharply during the 1980s. Meeting a moose in the wilds can be dangerous, but meeting one head-on in a car can be fatal. Even though signs warning "Moose Crossing" have been posted, moose-car collisions on northern New England roads are posing an increasing hazard for motorists (and for the moose). The table below provides data on moose-car collisions that occurred from 1980 to 1990. [1]

	1980	1984	1985	1988	1990
Vermont	0	unknown[2]	12	unknown	41
New Hampshire	unknown	unknown	49	117	170
Maine	156	215	unknown	unknown	500

1. Consider first the data for Vermont. On a separate sheet of paper, plot the number of collisions versus the year. ("Number of collisions" will appear on the vertical axis; "year" will appear on the horizontal axis.) How many *more* collisions were there in 1990 than there were in 1985? This number represents the increase in the number of collisions over that entire five-year period. Translate this number into the **average rate of change** in the number of collisions *per year* for that period.

2. Now calculate the average rate of change in the number of collisions per year for the period 1980–1985.

3. Compare the average rate of change (collisions per year) for 1980–1985 and for 1985–1990. Did the rate increase, decrease, or remain the same for the two consecutive five-year periods?

[1] "Comeback for Moose in New England Leads to Road Hazard," *The New York Times*, June 3, 1991.
[2] The missing numbers are not available; adequate records may not have been kept for those years. Life is like that, and we have to deal with the information we've got.

4. Write the equation of the line that passes through the data points corresponding to the years 1980 and 1985. (You'll need to decide what x represents: perhaps "years since 1980." Be sure, though, that it corresponds to the way you have drawn your graph.) Sketch this line on your plot of the data. If this line had accurately represented the relationship between the number of collisions and the year, how many collisions would you have expected in 1990? What, then, would have been the average increase per year in numbers of collisions from 1985 to 1990?

5. Make another graph using the data for New Hampshire. Do you think the points lie on a line? Check by computing the average increase per year in numbers of collisions for 1985 to 1988 and then for 1988 to 1990. Are these rates constant?

6. Finally, plot the data for Maine. Draw a line that passes through the data points corresponding to the years 1980 and 1984. Draw a second line that passes through the data points corresponding to the years 1984 and 1990. Without doing any calculations, state what happens to the average rate of change in numbers of collisions. As time passes, does that rate increase, decrease, or remain constant? Explain how you can answer this question by simply looking at your graphs.

Lab 2: Galileo

I've Got a Feeling I'm Falling

PREPARATION

In the first lab, we deduced a relationship between the temperature measured in degrees Fahrenheit, denoted F, and the temperature measured in degrees Celsius, denoted C. The relationship is given by the equation

$$F = \frac{9}{5}C + 32$$

We called C the *independent variable* and F the *dependent variable* in the equation. Because we can use the equation to produce exactly one value of F for each value of the variable C, the equation is said to define F as a *function* of C.

Another famous function that models the physical world was discovered in the 17th century by the great Italian scientist Galileo. Until Galileo's experiments with falling bodies, many scientists accepted Aristotle's contention that heavy objects fall faster than light ones do. Galileo studied the motion by measuring the distance an object falls in a given amount of time. Translating his data to our own units of measure, we have the following table of values for his two variables: t, which gives the time, measured in seconds, that the object has been falling, and s, which gives the total distance the object falls, measured in feet.

t	s
0	0
1	16
2	64
3	144
4	256

Before conducting his experiments, Galileo postulated that the distance fallen was proportional to the time elapsed. Remember that s is **proportional** to t if $s = k \cdot t$, where k is some fixed constant (called the constant of proportionality). Do you think Galileo's data confirmed his hypothesis? Test your answer using the data presented in the table.

Use a piece of graph paper to plot the points corresponding to the values in the table. Since it seems natural to think of s, the distance fallen, as dependent on t, the time the object has been falling, let s be the dependent variable measured on the vertical axis, and let t be the independent variable measured on the horizontal axis.

Be sure to label both axes carefully! For any graph, proper labeling means specifying the following next to each axis:

- the quantity being measured (for example, time, area, or population)
- the units in which that quantity is measured (days, square meters, thousands of persons, and so on)

You also need to show a scale on each axis. The numbers in the scale should be equally spaced: the length of axis from $s = 0$ to $s = 10$, for instance, should be the same as the length from $s = 40$ to $s = 50$.

A scale that is not uniform in this way will distort the shape of the graph and tell the wrong story. Choose scales that you think exhibit the data well; perhaps you will use a different scale for the s-axis than the one you use for the t-axis.

Bring your graph with you to the lab. In Lab 2, you will follow in Galileo's footsteps by finding a function relating s and t that models the data in the table. You will also use your grapher to investigate how a change in the independent variable *time* produces a change in the dependent variable *distance.*

Finally, you will work with a function from the same family as Galileo's function, but in an entirely different context.

©1991 by Sidney Harris

THE GALILEO LAB

No one will be able to read the great book of the Universe
if he does not understand its language,
which is that of mathematics.

—Galileo

In the preparation for this lab, you read that Galileo first postulated that the dependent variable s was proportional to the independent variable t. Compare your assessment of Galileo's hypothesis with those of your lab partners to make sure you agree. Let's use the graph you prepared for this lab to check your answer. You have learned that, if one variable is proportional to another, the graph of their relationship is a straight line through the origin with slope k, the constant of proportionality. In any of your team's graphs of s, do the points appear to lie on a straight line?

Now let's try to reproduce Galileo's work and guess the relationship between s and t. We need an algebraic expression in t that turns 1 into 16, 2 into 64, 3 into 144, and so on. That is, when t is 1, the algebraic expression has the value 16; when t is 2, it has the value 64. In function notation, we want to write

$$s = f(t)$$

where $f(0) = 0$, $f(1) = 16$, $f(2) = 64$, and so on. (Hint: All values of s in the table—including zero!—are multiples of the same number.)

What is the algebraic relationship between s and t?

The *domain* of a function is the set of allowable values for the independent variable. The *range* of the function is the corresponding set of values for the dependent variable. Although our table includes only a few integer values of t, you can use the formula for s as a function of t that you deduced above to compute values not in the table. Compute $f(\frac{5}{2})$.

We think of time as continuous; that is, to get from $t = 1$ to $t = 2$, we pass through all intermediate values of t. In particular, since $\sqrt{2}$ seconds is between 1 second and 2 seconds ($1 < \sqrt{2} < 2$), there must be a value of s corresponding to $t = \sqrt{2}$. What is it?

If you have not already done so, indicate that time is a continuous variable by connecting the points you plotted on your hand-drawn graph.

Do negative values of t make sense in this physical context?

In fact, Galileo's *model* for describing the fall of an object from a stationary position works well only near the surface of the earth (from the roof of a tall building, for example). Decide upon a reasonable set of values for the domain and range of Galileo's function, explaining your reasons for choosing those values.

(This task might seem difficult, because there is not one single correct answer. You need to use the information here and your common sense, and to make sure that the domain and range you give correspond to each other.)

Now use your grapher to draw the function you wrote to model the data in the table. You may need to rename the independent variable x and the dependent variable y to agree with the conventions of your particular grapher.

The graph on your calculator or computer screen might look very different from the one you drew by hand. There are two reasons for this.

The first reason has to do with horizontal and vertical scales. In the Fahrenheit lab, we saw that a change in scale on one axis can affect the appearance of the graph. One way to make the graph on the screen match the one you drew is to adjust the viewing window so that you see the same portion of the graph and that the scales match yours.

A more important difference is that the graph you drew represented Galileo's *model:* the values for t and s were chosen to make sense in the physical situation that the function modeled. The grapher, though, treats your function *abstractly*. It places only those restrictions on the domain and range that are a result of the algebraic operations performed on the independent variable. Here are two examples of algebraic restrictions: a computer or calculator (or you!) cannot evaluate $f(x) = 1/x$ at $x = 0$, nor can it find a real-number value corresponding to $g(x) = \sqrt{x}$ when $x = -5$. (Do you understand why?) Do you see any *algebraic* restrictions on the input to Galileo's function?

Because there is no algebraic reason for t not to take on any real-number value, the grapher will show much more of the graph of the function s than would make sense in the model.

Now use your grapher to help you determine the domain and range of the *abstract* function relating s and t that you—and Galileo—found. Vary the viewing window to see whether there appear to be any restrictions on either variable.

On graph paper, draw the portion of the graph of the abstract function on the interval from $t = -6$ to $t = 6$. With a second color, trace the part of the graph that models the falling-body problem. Use your graph to answer the following:

- When t changes from 1 to 2, what is the resultant change in s?
- When t changes from 2 to 3, what is the resultant change in s?
- When t changes from 3 to 4, what is the resultant change in s?

In the Fahrenheit lab, you saw that a fixed amount of change in the independent variable C produced a constant change in the dependent variable F. Does a fixed amount of change in the values of the independent variable t result in a constant change in the values of the dependent variable s? How do you think this difference between the Fahrenheit function and the Galileo function is reflected in their graphs?

For positive t, as t increases, does s increase or decrease? When does s increase more rapidly: when t increases from 1 to 2 or when t increases from 4 to 5? The change in s per unit change in t is called the **average rate of change** of s. Do you think that the average rate of change of s from $t = 9$ to $t = 10$ seconds will be greater or less than the average rate of change of s from $t = 2$ to $t = 3$?

Use your grapher to estimate the time at which the object will have fallen 100 feet. At what time will the object have fallen 500 feet?

JACK'S FIELD

The curve on your grapher is a **parabola**. It is the familiar and beautiful shape traced out by the trajectory of an arrow. Functions whose graphs are parabolas arise in many different situations.

Here's one: Suppose Jack has 128 feet of fencing to make a rectangular enclosure for his cow. He could fence in a rectangle 10 feet by 54 feet $(10 + 10 + 54 + 54 = 128)$. What would be its area? He could also fence in an area 24 feet by 40 feet. (Check that such a field would require 128 feet of fencing.) What is its area?

The distance around the border of the fenced region is called its **perimeter**. Each of the two fields we've just considered has a perimeter of 128 feet, but their areas are different. Draw a sketch of a rectangular field and label one side x. How could you represent the adjacent side of the field in terms of x so that the perimeter is 128 feet? Label the other three sides of your diagram and make sure that the perimeter is 128 feet.

Write an expression for the area, A, of the fenced region, measured in square feet, in terms of the length and width as expressions in x. Once a particular width x is chosen, A is completely determined, so x is the independent variable, and $A = f(x)$ is the dependent variable.

Now, here comes the parabola part. Use your grapher to sketch $A = f(x)$, then use the graph to help you answer the following questions. (You might have to change the viewing window.)

- What set of values for x make sense in this context? Explain how you determined your answer.
- What is the corresponding set of values for the dependent variable A; that is, what is the range of this function, used as a model for Jack's field?
- If Jack wants to enclose the largest possible rectangular area, how should he fence in his field; that is, what dimensions should he use? Describe how you used the graph to find the answer.

THE LAB REPORT

Your lab report should include two pairs of graphs:

- the abstract function $s(t)$ and Galileo's model for falling bodies
- the abstract function $A(x)$ and the area model for Jack's field

(Color coding and careful annotation can allow you to combine each model and its abstract function in a single graph.) Give the domain and range of each of the four functions and explain your answers.

Note the similarities and differences between the graphs of the abstract functions $A(x)$ and $s(t)$. Compare the formula for $A(x)$ to the formula for $s(t)$. What features account for the similarities in the graphs? What features account for the differences?

Contrast the rate of change of the dependent variable F in the Fahrenheit lab with the rate of change of the dependent variable s that you examined in this lab. What can you tell about the rates of change for s and F by looking at their *graphs*? What can you tell about them by looking at their *algebraic formulas*?

Finally, give the dimensions (length and width) of the rectangular field of largest area that Jack can enclose. Explain how you used the graph of A to determine your answer.

Section 2—Projects and Explorations

2.1: THROWING A CURVE—Exploring Quadratic Functions

What can a grapher help you to discover about quadratic functions? This exercise is designed to encourage you to explore the various possibilities for quadratics. Although you will probably not be able to find exact or complete answers in every case, you will nevertheless be able to use the rapid graphing capability of your computer or calculator to learn much more about quadratic functions than you could if you depended solely upon point plotting.

The general expression of a quadratic function takes the form

$$f(x) = ax^2 + bx + c$$

where a can be any value other than zero, and b and c can be any real numbers. You've already worked with quadratic functions such as $f(x) = 16x^2$ ($a = 16, b = 0, c = 0$) and $f(x) = -x^2 + 64x$ ($a = -1, b = 64, c = 0$).

1. Choose several other sets of values for a, b, and c and view the quadratics with your grapher. Sketch the basic shape(s) that characterize the graphs of quadratic functions. (Note that you are *not* asked to provide actual graphs, but rather to produce rough sketches of the shapes you see on the screen.)

What features of the graph of a quadratic function $f(x) = ax^2 + bx + c$ are controlled by the value of a? Of b? Of c? To answer these questions, you will need to be more methodical in your investigation, changing only one constant at a time.

2. Select a particular quadratic function by choosing any set of (nonzero) values for a, b, and c. Overlay several graphs: your chosen quadratic function and that function modified by different values for a.
   (a) How does changing a change the graph?

   (b) What happens to the graph if you change the *sign* of a?

(c) What happens if you keep the sign of a fixed, but increase the *magnitude* of a?

(d) As you increase the size of a, is there any point that doesn't change?

3. Repeat the steps in (2) to investigate what the coefficient b controls in the graph. The effect of b is more complicated than that of a. You probably won't be able to come to a definitive conclusion, but you should be able to see the difference between using positive values and negative values. If $b = 0$, what do you observe about the graph?

4. Now investigate c by trying several different values for that constant. You should be able to observe a specific effect that is easy to describe. Write a sentence or two summarizing what c does.

Mathematicians are very interested in knowing when (that is, for which x-values) a function is equal to zero (intercepts the x-axis), is positive (lies above the x-axis), or is negative (lies below the x-axis). Many "nicely behaved" functions can't switch sign without crossing the x-axis first.

5. Write a formula for a quadratic function that crosses the x-axis both at -2 and at 3. (You might want to begin by writing the quadratic in factored form.) On which interval(s) of the x-axis is your function positive? On which interval(s) is it negative? Illustrate your answer with a quick sketch of the graph.

6. Write a quadratic function whose graph *doesn't* cross the x-axis. If you attempt to use the quadratic formula to find x-intercepts for this function, what happens? Does the graph lie above or below the x-axis?

7. Write a quadratic function whose graph *touches* the x-axis but does not cross it. What is special about the factored form of this quadratic?

8. Is it possible for the graph of a quadratic function to cross the x-axis in more than two places? Explain why or why not.

2.2: DON'T FENCE ME IN

This activity is a follow-up to Lab 2

1. Recall Jack's field from the Galileo lab. Now suppose Jack has only 90 feet of fencing instead of 128 feet. Following the procedure you used in the lab, write a function for the area A of a rectangular field of perimeter 90 feet in terms of its width x. Use the graph of the function $A(x)$ to determine the dimensions of the rectangular field of greatest area that he can enclose. Does the *shape* of the largest rectangular field depend upon the amount of fencing? Explain.

2. In the space below, show a labeled sketch of the field and a graph of the area function, indicating the point on the graph that gave you both the "best" value for x (that is, the one that yielded the greatest area) and the maximum value for the area.

3. Jill, a mathematics student, decides to use the same amount of fencing as Jack (90 feet) to fence in a circular grazing field for the cow. Recall that the **circumference** (perimeter) of a circle is related to its radius r by the formula $C = 2\pi r$, and that the area A is also a function of r: $A = \pi r^2$. Use these two relationships and your calculator to find the area of Jill's field. What is the largest area Jack was able to obtain with 90 feet of fencing? Whose field is larger?

4. If a circular field has more area than a rectangular one, given a fixed amount of fence, why do you suppose farmers still lay out rectangular fields?

2.3: FEELING RUN DOWN?

Too much or too little air pressure can adversely affect the life of an automobile tire, so the owner's manual of a car gives the recommended pressure. To come up with these recommendations, a manufacturer performs a quality control test to measure the life of a tire for various amounts of air pressure. The model

$$y = -1106 + 69.7x - 1.06x^2$$

comes from one such study for a particular type of tire. The dependent variable y represents the life of the tire (measured in thousands of miles driven), and x represents tire pressure (measured in pounds per square inch, or *psi*).

1. Sketch a graph of the abstract function.

2. A reasonable domain for the model is $28 \leq x \leq 38$ psi, because all of the data fell within that interval. In a contrasting color, show the portion of the graph that represents the model.
3. What is the range of the model?

4. What is the optimal tire pressure? How many miles do we expect the tire to last if its pressure is kept at the optimal level? (Pay attention to the units when answering this question.)

5. Compare the average rate of change in tire life when the pressure increases from 28 to 30 psi with the average rate of change when the pressure increases from 30 to 32 psi. If you were driving around on underinflated tires, in which of the two cases would the extra two psi be more important to you? Explain.

2.4: DIAGONALS

God ever geometrizes.

— Plato

In this project, you'll investigate a surprising property of some very common geometric objects. The figures below are polygons (from the Greek *poly*, many, and *gon*, angle). The first two polygons below are **convex** (they have no dips or dents); the third is nonconvex. Polygons have their own names reflecting the number of angles they have. Thus, the first polygon below is called a pentagon, since the prefix *penta* is the Greek name for five, and the second is called a heptagon, because *hepta* means seven.

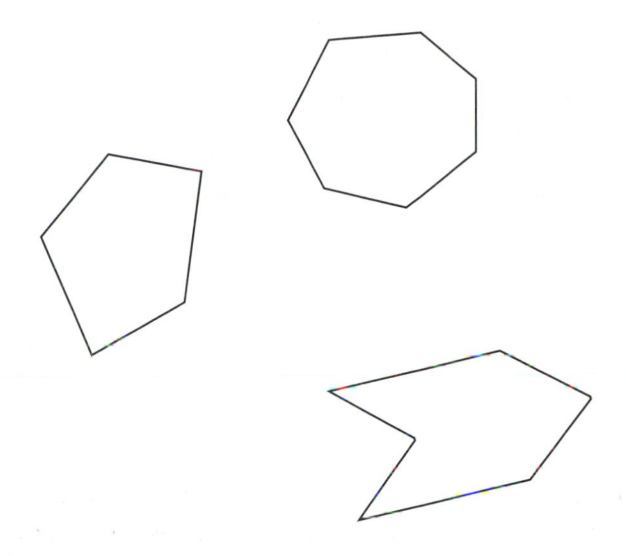

1. A hexagon is a polygon with six angles. Above, draw two different convex hexagons.
2. A **diagonal** of a polygon is a line connecting two nonadjacent vertices or angles. Draw all possible diagonals on each of the two convex polygons above and on your two home-grown hexagons. (Did you get nine diagonals for each hexagon?)
3. In fact, the number of diagonals of a convex polygon depends only upon the number of its angles; that is, if n is the number of angles of a convex polygon and D is the number of its diagonals, then D is a function of n. (We use n instead of x for the independent variable to indicate that the only allowable values for n are integers.) Fill in just the second column of the table, giving the values of the dependent variable D corresponding to each of the values of the independent variable n. You'll use the remaining two columns later.

n	D	First differences	Second differences
3			—
4			
5			
6			
7			
8			
9			—

As we did for the functions in the Fahrenheit and Galileo labs, let's try to determine a formula for D.

4. Could D be a linear function of n? Give numerical evidence for your answer.

5. Here's a way to double-check your answer to (4). Notice that each n-value is separated from the next by one unit. If D were a linear function, a one-unit change in n would always result in the same amount of change in D. In the third column of the table, enter the difference between successive D-values (subtracting each one from the one that follows it). Are these first differences constant?

6. Using the table to find a formula for D in terms of n is tricky. It would help if we knew what kind of function D is. Here's one way to find out. In your table, fill in the last column with successive second differences by subtracting each value in the first-differences column from the one following it. What do you observe?

7. The phenomenon that you should note in (6) tells a mathematician that D is a *quadratic* function of n. Convince yourself of this by plotting the ordered pairs (n, D) from the table, either on graph paper or on your grapher, and noticing that they appear to lie along the right half of a parabola. But which parabola? We're about to find out.

8. One allowable value of n gives a D-value of 0. Which is it? This value of n is called a **root**, or n-intercept, of D, so we know that ($n -$ this value) must be a factor of the quadratic expression for D. In other words, if r is the root you found, then D is the product of $(n - r)$ and some other linear factor. Experiment with the numbers in your table until you discover the other linear factor of D, and write the function equation that gives D as a function of n.

9. Using the formula you found for D, what is the number of diagonals of a convex polygon with 25 angles?

10. Solve a quadratic equation to determine the number of angles in a convex polygon with 104 diagonals.

11. The variables n and D in this mathematical model are called **discrete**, because their values are separated by intervals. Write the domain of the model (the set of numbers that make sense as inputs), remembering that it is a discrete set.

Lab 3: Graph Trek Explorations

To boldly go where no lab has gone before

PREPARATION

Before you begin this lab, let's be sure that you understand function notation. Given a specific formula for a function, say,

$$f(x) = 3x^2 - 2x + 5,$$

write the formulas for the following:

$$f(x) + 1, \quad f(x + 1), \quad -f(x), \quad f(-x), \quad f(2x), \quad 2f(x)$$

Repeat this exercise, changing the formula for $f(x)$ to $3\sqrt{x - 1}$. Check your answers with the ones given at the end of this preparation.

Function notation is very efficient because it allows us to express the algebraic rule for a mathematical function economically, using only a few symbols. This very economy of expression, unfortunately, can also be a source of confusion. Do you know, for example, the difference in meaning between $f(x) + c$ and $f(x + c)$, or between $-f(x)$ and $f(-x)$?

In the Galileo lab, you studied the quadratic function $16x^2$. You probably plotted several points in order to draw its graph. Suppose that function were modified slightly so that you had to draw $16(x + 3)^2$ or $100 - 16x^2$. Would you want to plot a whole new set of points? Think of all the work! In this lab, you will discover quick methods for graphing functions such as these without point plotting.

In Lab 3, you will imagine yourself a research mathematician. The purpose of the lab is to understand what happens to the graph of any function $f(x)$ when you apply a **transformation** to the function—that is, when you change its algebraic formula in one of the ways you did above. You will have three research tasks during the lab:

- collecting information from your grapher
- deciding the effect of the given transformation (backed up by examples)
- determining why the effect occurred

As background for this research, you will need to be familiar with the general shapes of the following functions. Sketch their graphs below.

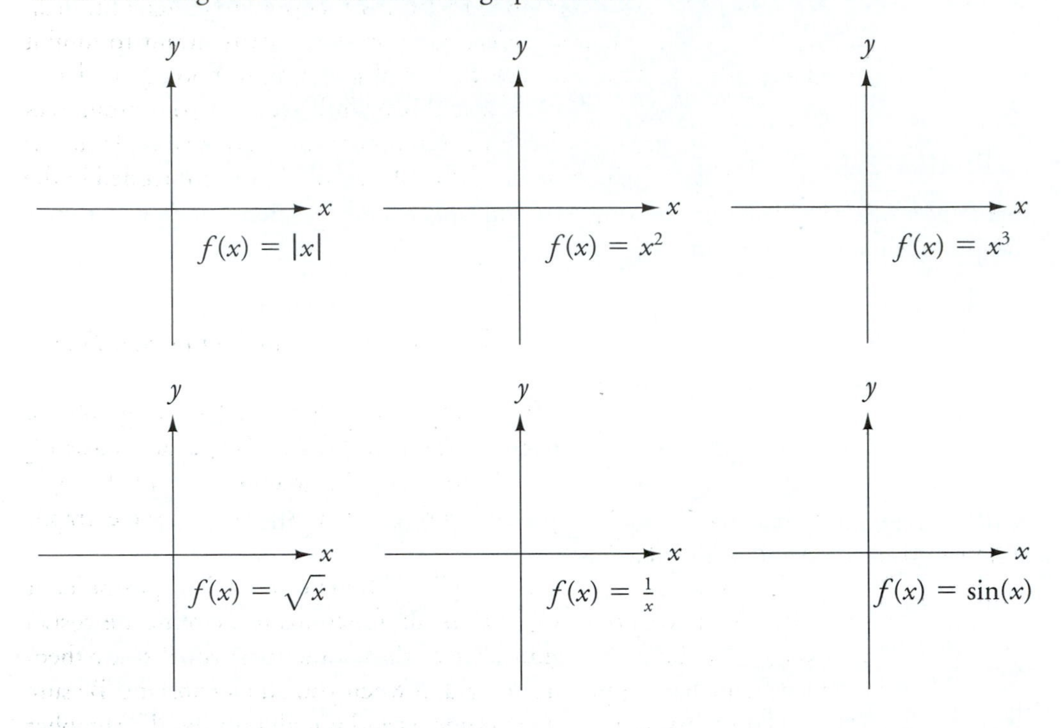

Answers for the Preparation Exercises

If	$f(x)$	is	$3x^2 - 2x + 5,$	If	$f(x)$	is	$3\sqrt{x} - 1,$
then	$f(x) + 1$	is	$3x^2 - 2x + 6$	then	$f(x) + 1$	is	$3\sqrt{x - 1} + 1$
	$f(x + 1)$	is	$3(x + 1)^2 - 2(x + 1) + 5$		$f(x + 1)$	is	$3\sqrt{x}$
	$-f(x)$	is	$-3x^2 + 2x - 5$		$-f(x)$	is	$-3\sqrt{x - 1}$
	$f(-x)$	is	$3x^2 + 2x + 5$		$f(-x)$	is	$3\sqrt{-x - 1}$
	$f(2x)$	is	$12x^2 - 4x + 5$		$f(2x)$	is	$3\sqrt{2x - 1}$
	$2f(x)$	is	$6x^2 - 4x + 10$		$2f(x)$	is	$6\sqrt{x - 1}$

THE GRAPH TREK LAB

Pay particular attention to the title of this lab. The lab instructions are designed to help you discover and understand some mathematical patterns. They aren't meant to inhibit your own creative investigations, however. The beauty of graphing technology is that it allows us to try out our ideas quickly. ("What would happen if . . . ?") If your group sees something interesting or puzzling, discuss it with one another. The individual explorations of each lab group are an important component of this lab and ought to be included in the report.

Shifts

The first transformation to examine is that of adding a constant c to the function $f(x)$—that is, graphing $f(x) + c$.

Start by graphing the simplest parabola, $f(x) = x^2$. Now overlay the graphs of $y = x^2 + 3$ and $y = x^2 - \sqrt{2}$. Describe what you see; that is, when $f(x)$ is defined as x^2, how does the graph of $y = f(x) + 3$ compare? How does the graph of $y = f(x) - \sqrt{2}$ compare? You should be able to make quick and fairly accurate sketches of these graphs without resorting to point plotting.

Change the formula for $f(x)$. You might use a linear function, another quadratic, or something totally different. Two particularly interesting functions to examine are $\cos(x)$ and $\sin(x)$. (Don't worry yet about the significance of those functions; you'll study them in detail later.) Now watch what happens to the graph when you add a constant. Be sure to use several different constants: not all whole numbers and not all positive. The grapher is nothing but a very fast point plotter, and it calculates the points for each new graph without reference to the previous one. *You*, however, can recognize a pattern that enables you to get a new graph from the old one with a minimum of effort. Do several different examples until you are certain you understand the pattern. Describe, in general terms, what happens to the graph of a function when you add a constant to the function.

Suppose the graph of some function $y = CUP(t)$ looked like this:

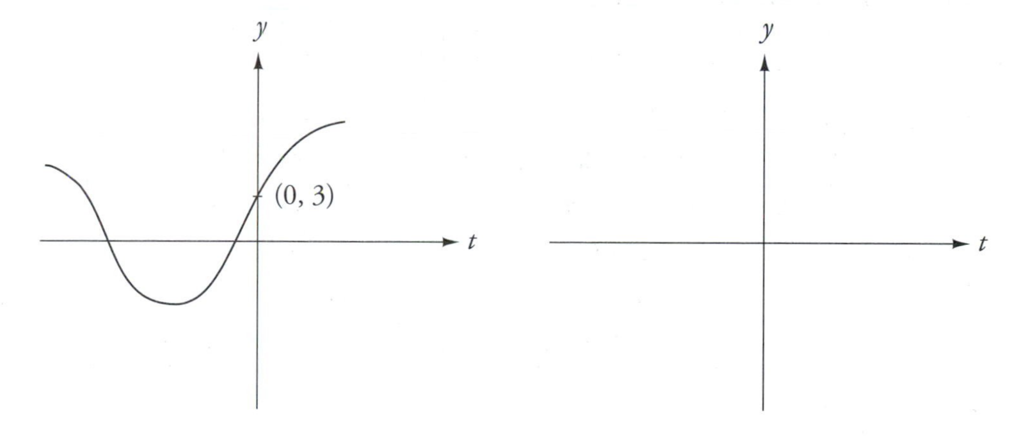

Draw the graph of $y = CUP(t) - 4$. Label its y-intercept.

So far, you've been adding a constant to the function, the dependent variable. Suppose, instead, you were to add a constant to the independent variable; that is, you first add the constant to the input variable, and then perform the function. What do you think would happen to the original graph? From your work in the preparation section, you know that the order of operations makes a difference in the *formula* for the function. (Compare the formulas for $f(x) + 1$ and $f(x + 1)$.) Now you will see that the order of operations also makes a difference in the *graph* of the function.

What do you expect the graph of $y = (x + c)^2$ to look like? (What will be its relationship to the graph of $y = x^2$?) Make your prediction.

Now, graph $f(x) = x^2$ and overlay the graph of $g(x) = f(x + c)$, that is, of $(x + c)^2$. (You choose a value for c.) Is this what you expected? Explain *what* happened and *why.* (The "why" is a lot harder than the "what," but don't avoid thinking about it.) How does your choice of a value for c affect the basic graph of x^2?

Test your conjecture by trying several different values for c, both positive and negative values.

Do you think the behavior you've witnessed is something peculiar to parabolas, or will other functions respond in the same way? Check out your intuition by changing the function. Here are some suggested functions: x^3, $\sqrt{x}$, $\sin(x)$, and $|x|$. Try several examples of adding a constant to the independent variable *before* performing the operation, and describe the effect on the graph. (You'll be looking at graphs of such functions as $\sqrt{x - \frac{1}{2}}$ and $|x + \pi|$ and comparing them to the graphs of the simpler functions they most resemble, $\sqrt{x}$ and $|x|$ in this case.)

Suppose you start with $f(x) = x^2 + x$. Adding a constant to the independent variable means looking at, say, $f(x - 1)$. What is the algebraic formula for $f(x - 1)$? How would its graph be related to that of $f(x)$?

Now generalize about the relationship between the graph of some function $BALE(a)$ and that of $BALE(a + c)$. Suppose the graph of $y = BALE(a)$ looks like this:

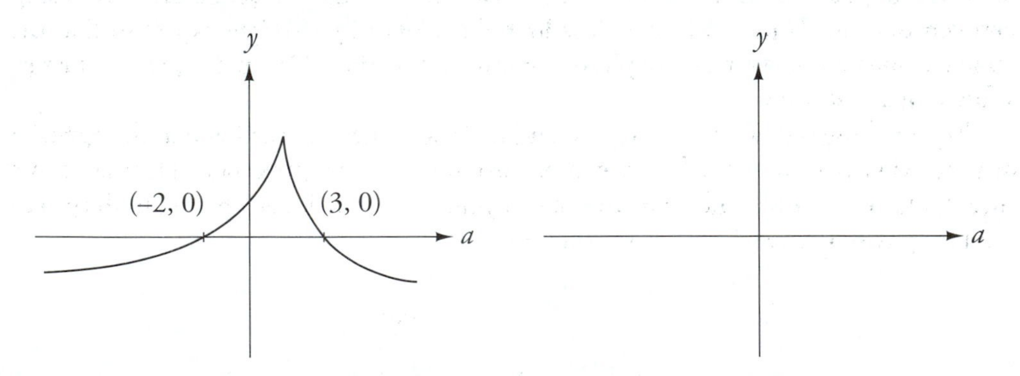

Draw the graph of $y = BALE(a + 2)$. Label its a-intercepts.

By now you should be able to look at the formula for a function and tell whether its graph involves a shift of one of the basic shapes you've been learning. You should be able to predict how far and in what direction(s) the basic shape is shifted. For each of the following, identify the basic function, predict the kind(s) of shift, then have the grapher draw the graph.

(a) $\sqrt{x+2}$

(b) $\sin(x-1)$

(c) $(x+2)^3$

(d) $(x-3)^2-5$

(e) x^3+2

(f) $1+\sin(x)$

The function $|x|$ and its relatives have V-shaped graphs. Be sure your interval allows you to see both sides of the V.

(g) $2+|x-1|$

The last two graphs each have two branches. Choose an interval that shows the whole picture.

(h) $\frac{1}{x}-3$

(i) $\frac{1}{x-3}$

REFLECTIONS

Another way to get new graphs from old is through reflections. Sketch the graph of $y=-x^2$. (Note that $-x^2$ is different from $(-x)^2$. Check it quickly if you aren't sure.) With the graphs of x^2 and $-x^2$ on the screen at the same time, describe the relationship between the two. Since mathematicians look for patterns, you might expect to find the same relationship between the graphs of any pair of functions $f(x)$ and $-f(x)$ — that is, a function and its opposite.

Try graphing several other functions and their opposites. Remember that the opposite of a function is simply the function itself preceded by a negative sign. Here are some suggestions, but you should make up a few of your own as well. Each person in the group should try two or three. Compare your results.

$\sqrt{x}$ $\sin(x)$

x^2+3x-4 (Remember to take the opposite of the entire quadratic!)

Now it's time to generalize. Suppose the graph of some function $y = SICK(u)$ looks like this:

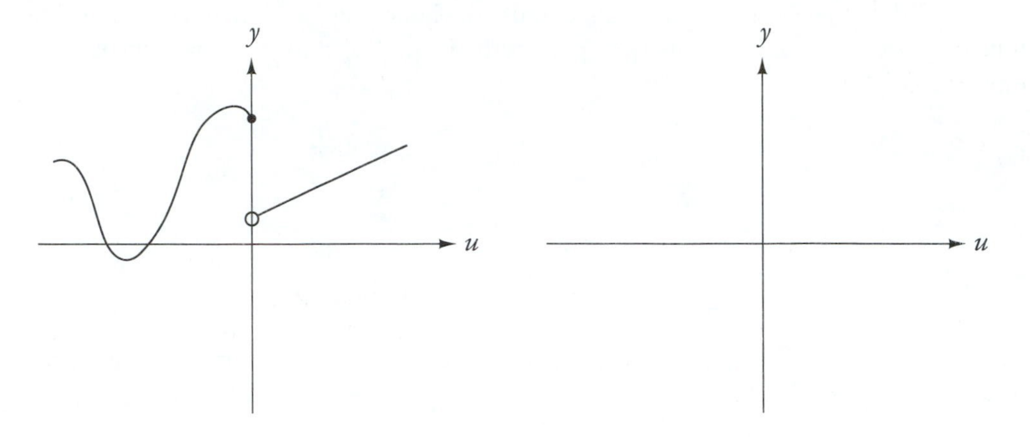

Sketch the graph of $y = -SICK(u)$.

You have been comparing the graph of a function to that of its opposite. Suppose you look instead at $f(-x)$, that is, at the original function when the input variable is replaced by its opposite. Try this with the function $f(x) = \sqrt{x}$. What's the formula for $f(-x)$? What's the relationship between the graphs of $f(x)$ and $f(-x)$? If you don't see two separate graphs on the screen, pick a different window. Compare these two graphs with the graph of $-\sqrt{x}$. Try to get all three graphs on the screen simultaneously. The position of the negative sign makes a difference! Explain.

You should now see one graph in the first quadrant, one in the second, and one in the fourth. Can you complete a symmetrical pattern by putting a graph into the third quadrant as well? You'll have to figure out its algebraic formula first.

Try graphing $f(-x)$ for several other functions, perhaps the same ones you used for learning about $-f(x)$. Keep doing examples until you can predict with confidence what the graph of $f(-x)$ will look like.

What happens if you start with $f(x) = x^2$? Why?

Now sketch the graph of $y = SICK(-u)$.

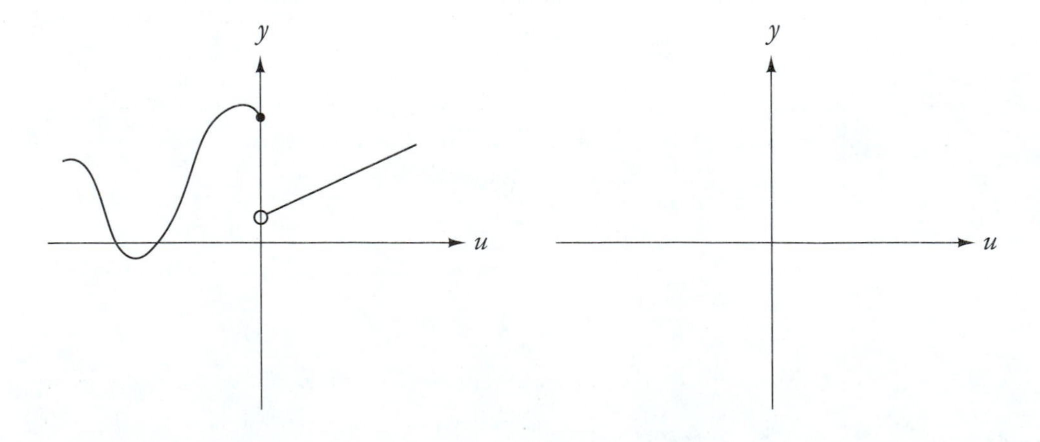

At this point, you should be able to draw the graphs of many different functions by sketching the simple function they most resemble and applying appropriate reflections and shifts. Try these:

$-\sqrt{x-2}$

$2-\sqrt{x}$

$2+\sqrt{-x}$

$5-(x+3)^2$

$|-x|-1$

$-\dfrac{1}{x+1}$

THE LAB REPORT

What happens to the graph of a function when a constant is added to the output or to the input? What happens to the graph of a function when the output or the input is replaced by its opposite? Your report should answer these questions fully but concisely.

- Summarize what you learned about the graphs of $f(x)+c$, $f(x+c)$, $-f(x)$, and $f(-x)$ by describing how their graphs are related to the graph of $f(x)$. Tell the effect of each type of transformation and explain why each effect occurred.
- Illustrate with representative sketches of a few well chosen functions. You should not submit every single graph you drew during the lab, but be sure that the graphs that form part of your report are sufficiently general to make the patterns clear.
- Include any interesting or puzzling observations your group may have made.

Section 3—Projects and Explorations

3.1: FUN HOUSE MIRRORS—Vertical Stretching and Compression, $c \cdot f(x)$

This exploration is a follow-up to Lab 3

In this assignment, you'll look at another way to modify a function and watch what happens to its graph. In order to get accurate pictures, you should have equal scales on the horizontal and vertical axes. Adjust your grapher so that the x- and y-axes have the same scale (that is, use square scaling), and keep it that way for this entire assignment.

1. Sketch the graph of $\sin(x)$ on the interval $-3\pi \leq x \leq 3\pi$. Overlay the graph of $3\sin(x)$, and describe what the "3" does to the original graph.

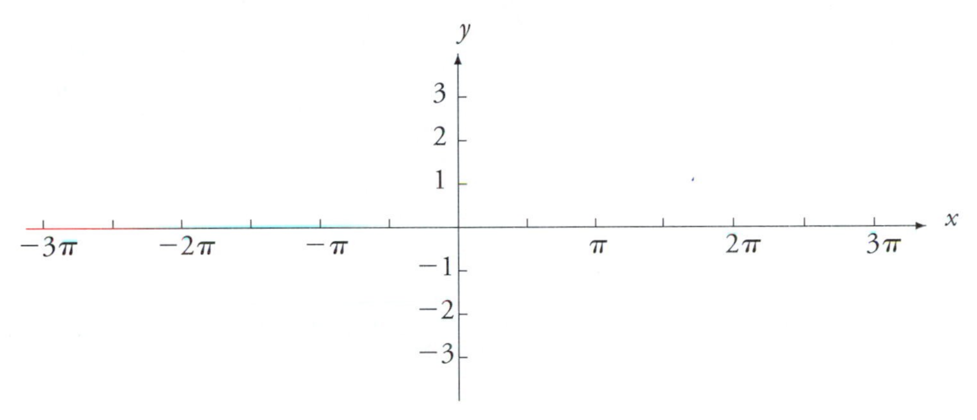

What do you observe about the height of each peak?

The original graph contains the point $(\frac{\pi}{6}, 0.5)$, or approximately $(0.52, 0.5)$. What point on the new graph has the same x-value?

What do you notice about the x-intercepts?

2. Now overlay the graph of $\frac{1}{2}\sin(x)$ and describe the effect of the "$\frac{1}{2}$." What features change, and in what way? What features remain the same?

Now you will examine $c \cdot f(x)$ for several other functions. (But don't feel that these are the *only* ones you may explore.)

3. Compare $3\sqrt{x}$ and $0.3\sqrt{x}$ to $\sqrt{x}$. Make sketches. Identify each graph.

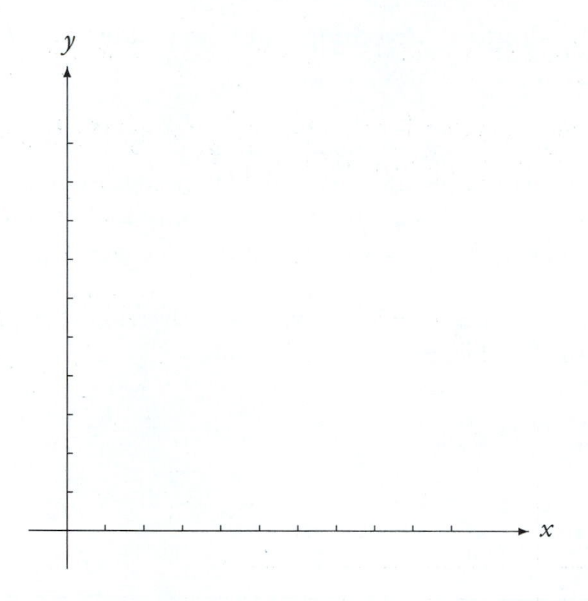

4. Compare $2\cos(x)$ and $\frac{1}{2}\cos(x)$ to $\cos(x)$. Make sketches. Describe what you see.

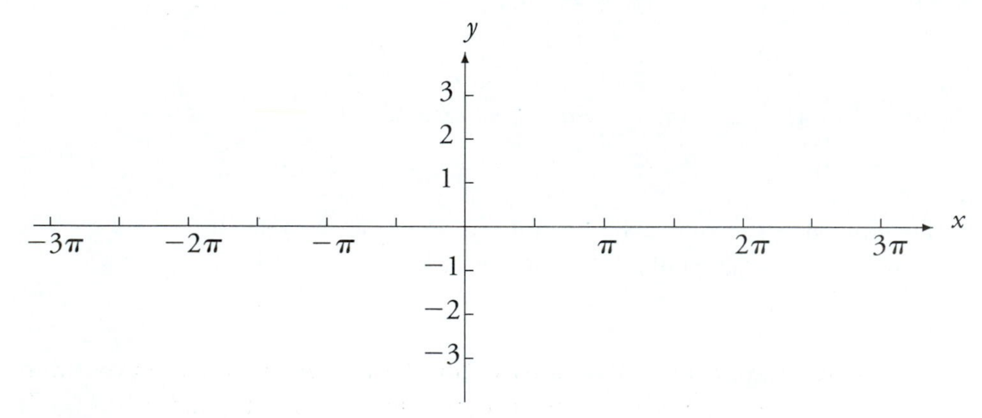

5. Do one more set of your own choosing.
   (a) Pick a function $f(x)$, one whose graph extends both above and below the x-axis.
   (b) Sketch its graph.
   (c) Sketch two more graphs, $a \cdot f(x)$ and $b \cdot f(x)$, where a and b are constants, $a > 1$ and $0 < b < 1$.
   (d) Write a couple of sentences to generalize the effect on the graph of a function of multiplying the function by a positive constant.

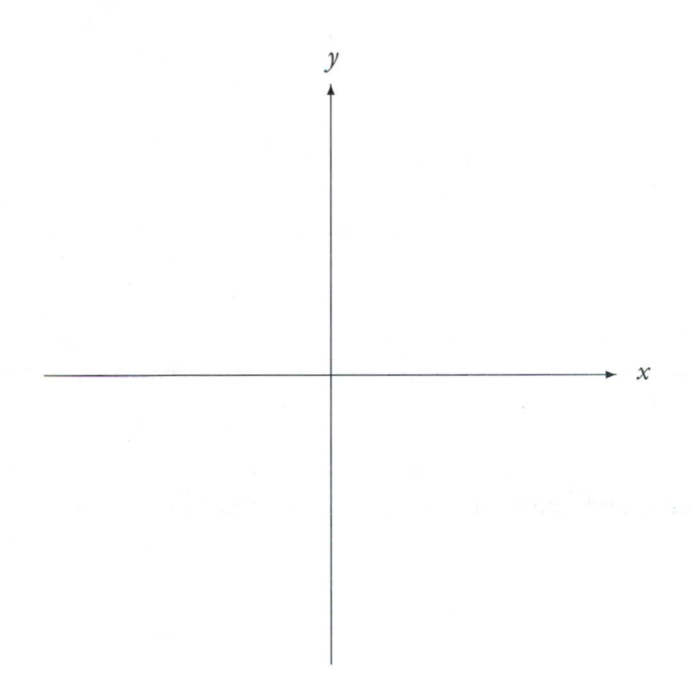

6. Now examine $c \cdot f(x)$ where c is negative. You already have an idea, from your work in Lab 3, of what might happen. Try a couple, such as $-3 \cdot f(x)$ and $-\frac{1}{2} \cdot f(x)$. In a sentence or two, generalize what you see.

3.2: THE ACCORDION EFFECT—Horizontal Stretching and Compression, $f(c \cdot x)$

This exploration is a follow-up to Lab 3

Here's yet another way to modify a function algebraically — multiplying the *independent* variable by a constant. The best functions to start with are $\cos(x)$ and $\sin(x)$, even though you might not yet understand their significance, because they clearly show the difference between multiplying an entire function by a constant (first perform the function, then multiply) and multiplying the independent variable by the constant (first multiply the input variable, then perform the function). Use square scaling, unless a different scale is indicated.

1. Recall, from the previous assignment, what happened to the graph of $\cos(x)$ when you changed it to $2\cos(x)$. Now try $\cos(2x)$. Show by means of a sketch that the "2" in $2\cos(x)$ *stretches* the graph of $\cos(x)$ vertically by a factor of 2, whereas the "2" in $\cos(2x)$ *compresses* the graph of $\cos(x)$ horizontally by a factor of 2.

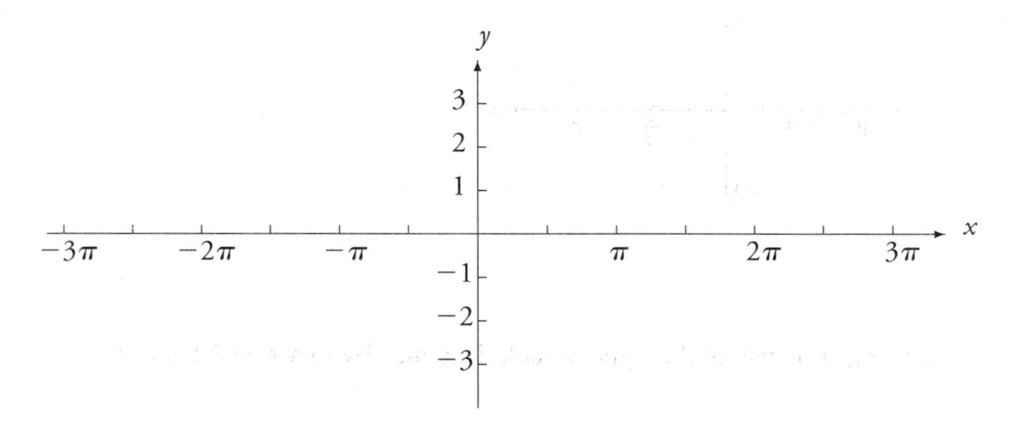

2. What would you expect of $\cos(\frac{1}{2}x)$? Graph $\frac{1}{2}\cos(x)$, $\cos(x)$, and $\cos(\frac{1}{2}x)$. Describe what you see.

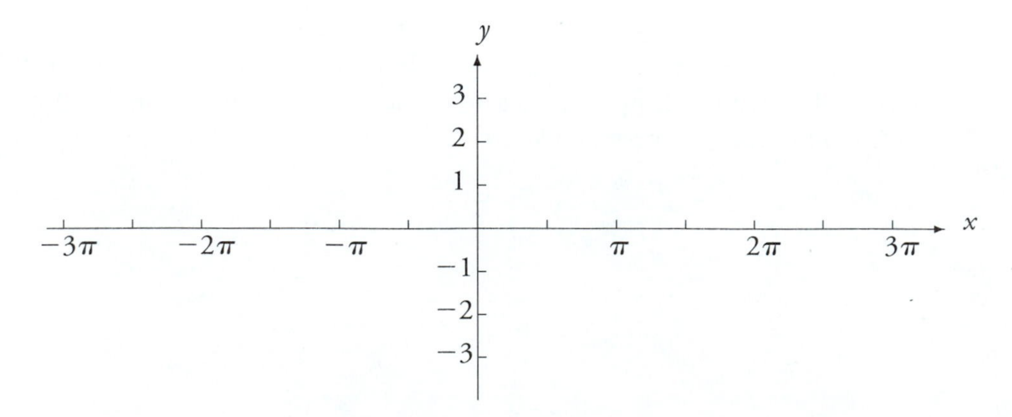

3. Scales make a difference! Draw each graph in the indicated window.

(a) $y = \cos(x)$

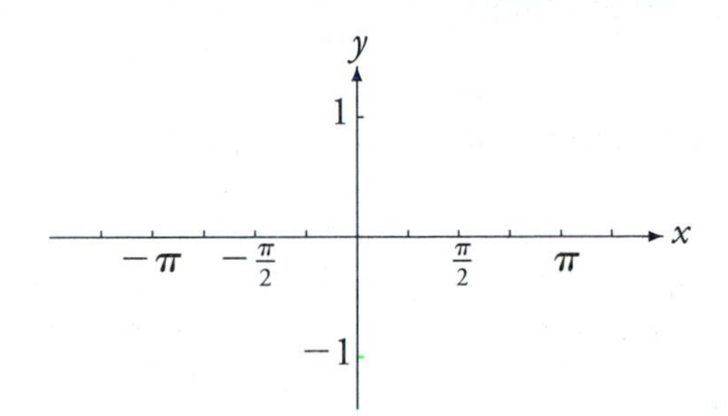

(b) $y = \cos(2x)$

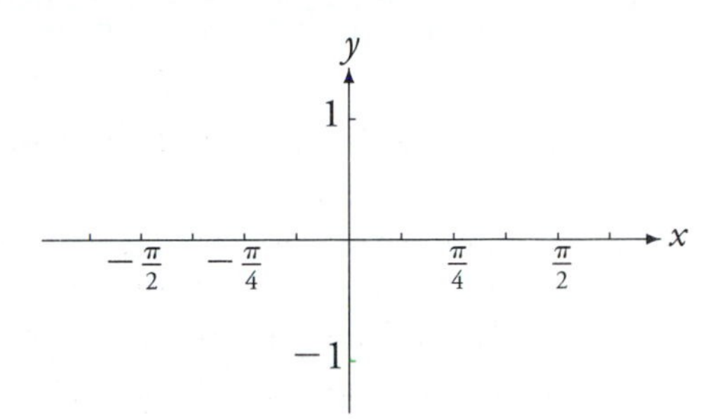

(c) $y = \cos(2x)$

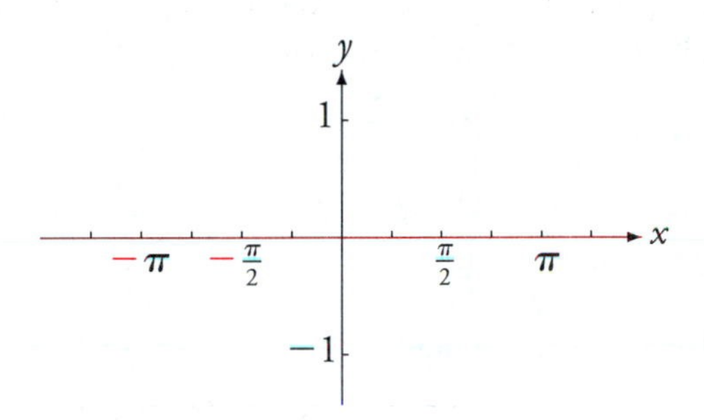

At first glance, two of these graphs look the same. Are they really? Explain.

4. Will the sine function behave in the same manner as the cosine function? Choose a value for c (something other than 2); graph $\sin(x)$, $c \cdot \sin(x)$, and $\sin(c \cdot x)$. Sketch what you see, using an interval wide enough to show complete waves and being careful to indicate the x-intercepts.

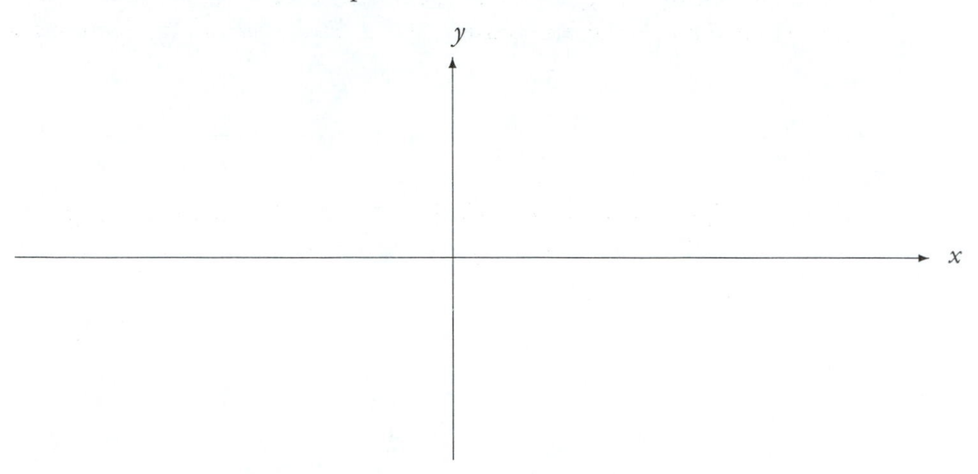

5. Let $f(x)$ be the function $x^3 - x$. Sketch its graph in the vicinity of its x-intercepts. On the same axes, sketch the graphs of $2f(x)$ and $f(2x)$. Provide a scale. (You may need to play around with the viewing window to get a good picture. Get close to the intercepts. Each graph has two loops; be sure you see them.) What effects do you see? Stretching or compression? Vertical or horizontal?

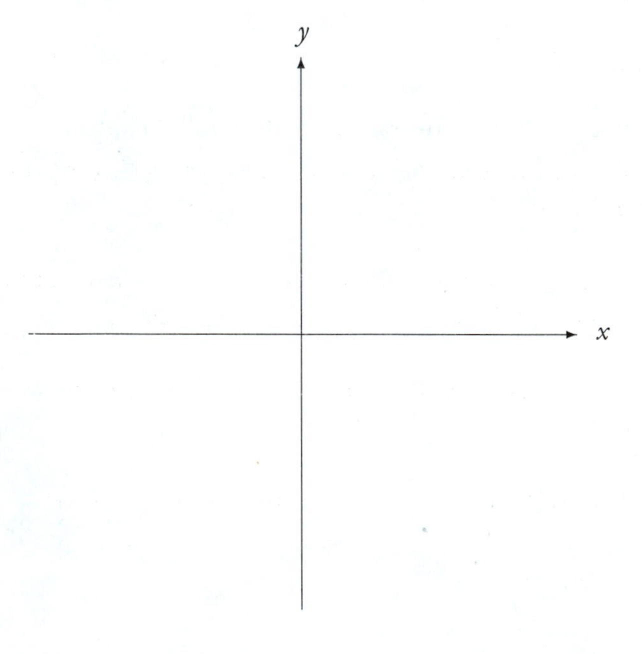

THE SILLY-PUTTY APPROACH TO GRAPHING

6. The effects you've been seeing apply to all functions, although it's often difficult to tell whether a graph has been stretched vertically or compressed horizontally. Take, for example, the function $f(x) = x^2$ and consider the vertical stretch $9 \cdot f(x)$. What value of c, in a horizontal compression $f(c \cdot x)$, will result in an identical graph?

7. Sometimes, a horizontal stretch can look like a vertical stretch. Consider the function $g(x) = \frac{1}{x}$ and a horizontal stretch, $g(\frac{1}{4}x)$. What vertical stretch $c \cdot g(x)$ will produce the same graph?

Multiplying a function by a constant has one effect on its graph; multiplying the independent variable by a constant has a different effect (even though, as you just saw, the results occasionally can be similar). After finishing this assignment and the preceding one, you should understand both effects.To check your progress, answer the following questions about these functions:

$$1 - \sin(0.2x), \qquad 60\sin\left(x - \frac{\pi}{4}\right), \qquad \frac{2}{5}\sin(x+1), \qquad \sin(3x) + 4$$

8. Which of the functions has a constant that behaves like a trash compactor, squashing the sine graph toward the x-axis?

9. Which has a constant that grabs the tops and bottoms of the sine wave and tugs them vertically, away from the x-axis, as in stretching a rubber sheet?

10. Which has a constant that stretches $\sin(x)$ horizontally away from the y-axis, as with an overextended accordion?

11. Which has a constant that squeezes the sine wave horizontally, toward the y-axis?

For further exploration, you might look at what happens to $f(c \cdot x)$ when the constant c is negative.

3.3: SELF REFLECTION—Symmetry about the y-axis ("Even" Symmetry)

This project is a follow-up to Lab 3

Beauty is truth, truth beauty,—that is all
Ye know on earth, and all ye need to know.
—John Keats, *"Ode to a Grecian Urn"*

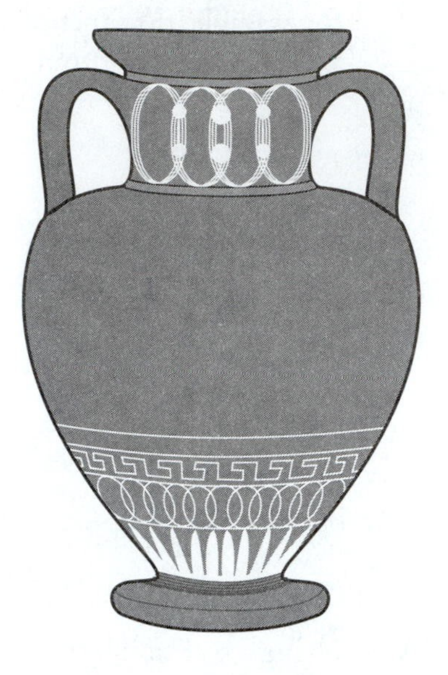

Beauty is often associated with symmetry. We would all agree that the shape of a Grecian urn is very symmetrical. When the curve representing the graph of a function $f(x)$ is placed symmetrically on the coordinate axes, we say it is **symmetric about the y-axis** because, if you folded this paper along the y-axis, the two halves of the graph would lie on top of each other. (Try it!)

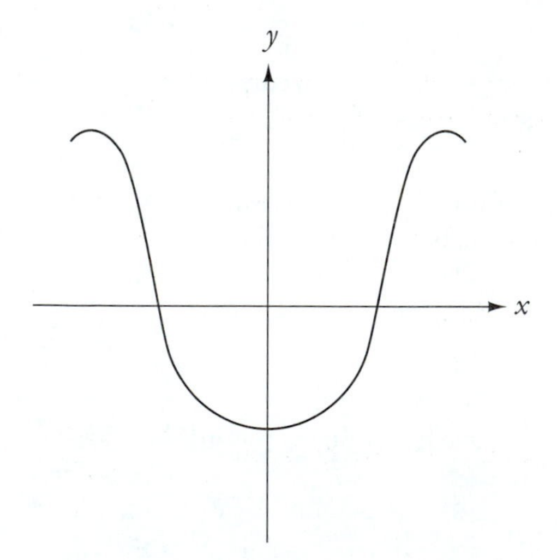

Looking at the graph of a function, you can probably decide whether or not it is symmetric about the y-axis. But can you tell from its algebraic formula alone? In this investigation, you will learn an algebraic technique for deciding, even without looking at the graph, whether or not a function is symmetric about the y-axis.

1. Here are the graphs of four more functions, $f(x)$, three of which are symmetric about the y-axis. Which three? Now, for each of these functions, draw the graph of $f(-x)$. (In the Graph Trek lab, you learned how to draw the graph of $f(-x)$, given the graph of $f(x)$. Recall that $f(-x)$ is *not* an upside-down version of $f(x)$!)

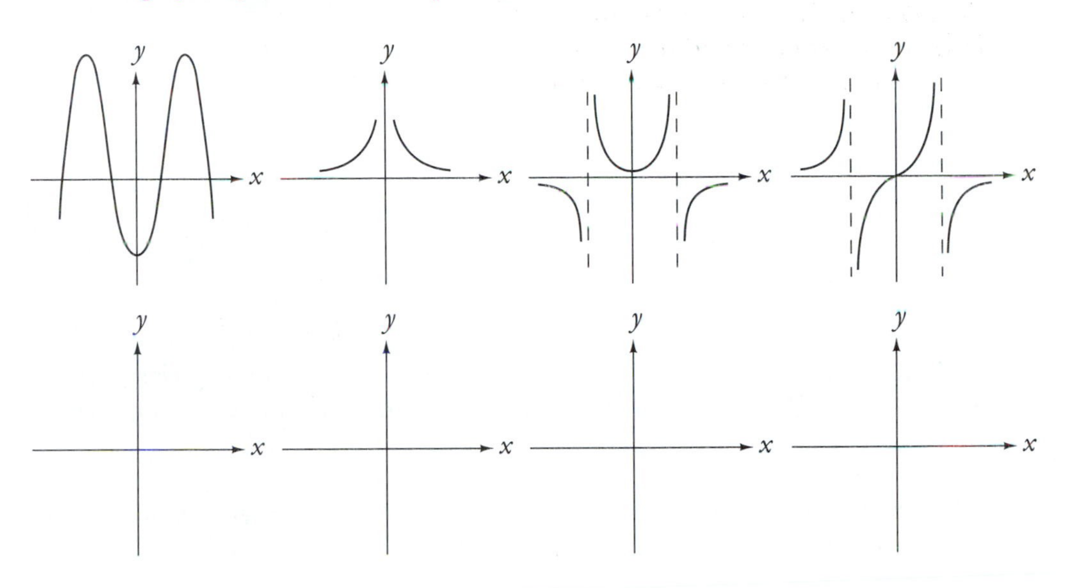

What do you observe about three of the graphs you just drew, something that isn't true of the other one?

2. Here are the formulas for the four functions pictured.
For each one, write the formula for $f(-x)$ and compare it to the formula for $f(x)$. What do you observe?
   (a) $f(x) = -x^4 + 10x^2 - 9$

   (b) $f(x) = \dfrac{1}{x^2}$

   (c) $f(x) = \dfrac{1}{4 - x^2}$ $\left(\text{Warning: remember, this really means } \dfrac{1}{4-(x^2)}.\right)$

   (d) $f(x) = \dfrac{x}{4 - x^2}$

3. Recall from the Graph Trek lab how the graphs of $f(x)$ and $f(-x)$ are related. In your own words, explain why you should expect the algebraic results you just obtained, simply by looking at the pictures.

In fact, the formula for a function reveals whether or not its graph will have y-axis symmetry. Whenever the formula for $f(-x)$ is the same as the formula for $f(x)$, the graph of the function will be symmetric about the y-axis.

4. Test the following algebraically to predict whether or not their graphs will be symmetric about the y-axis. Check your results by looking at their graphs.
   (a) $7 - 3x^2$

   (b) $x^6 - 40x^2 + 5$

   (c) $x^6 - 40x^3 + 5$

   (d) $\dfrac{x}{x^2 - 4}$

   (e) $\dfrac{x^2}{x^2 - 4}$

   (f) $\dfrac{x}{x^3 - 4x}$

   Simplify the fraction you write as much as possible.

3.4: PUTTING A SPIN ON IT—Symmetry about the Origin ("Odd" Symmetry)

This project is a follow-up to Lab 3

Spiral Galaxy NCG #2997 ©Anglo-Australian Observatory

The curve here mimics the symmetry of the galaxy above. It is placed on the coordinate axes symmetrically, but its symmetry is different from the even symmetry of the Grecian urn curve in the previous exercise. One way to describe its symmetry is to imagine placing a pin through the graph at the origin and then spinning or rotating the paper 180° around the pin, as a pinwheel (or a galaxy) rotates about its center. The new position of the graph is identical to its old position. We say such a graph is **symmetric about the origin**.

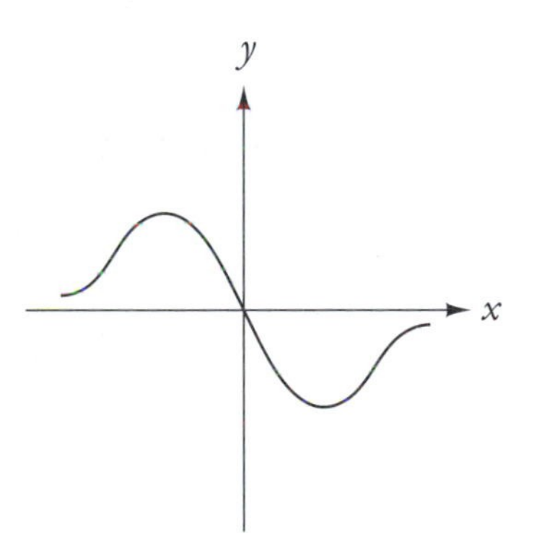

There is another algebraic technique for deciding from a function's formula whether or not the function has origin symmetry. Before learning that trick, though, you should perform several reflections.

1. Here are the graphs of four functions, $f(x)$, three of which are symmetric about the origin. Which three? Below each one, draw the graph of $f(-x)$. Below that, in the third row, draw the graph of $-f(x)$. (Be sure to use the *original* $f(x)$ each time.)

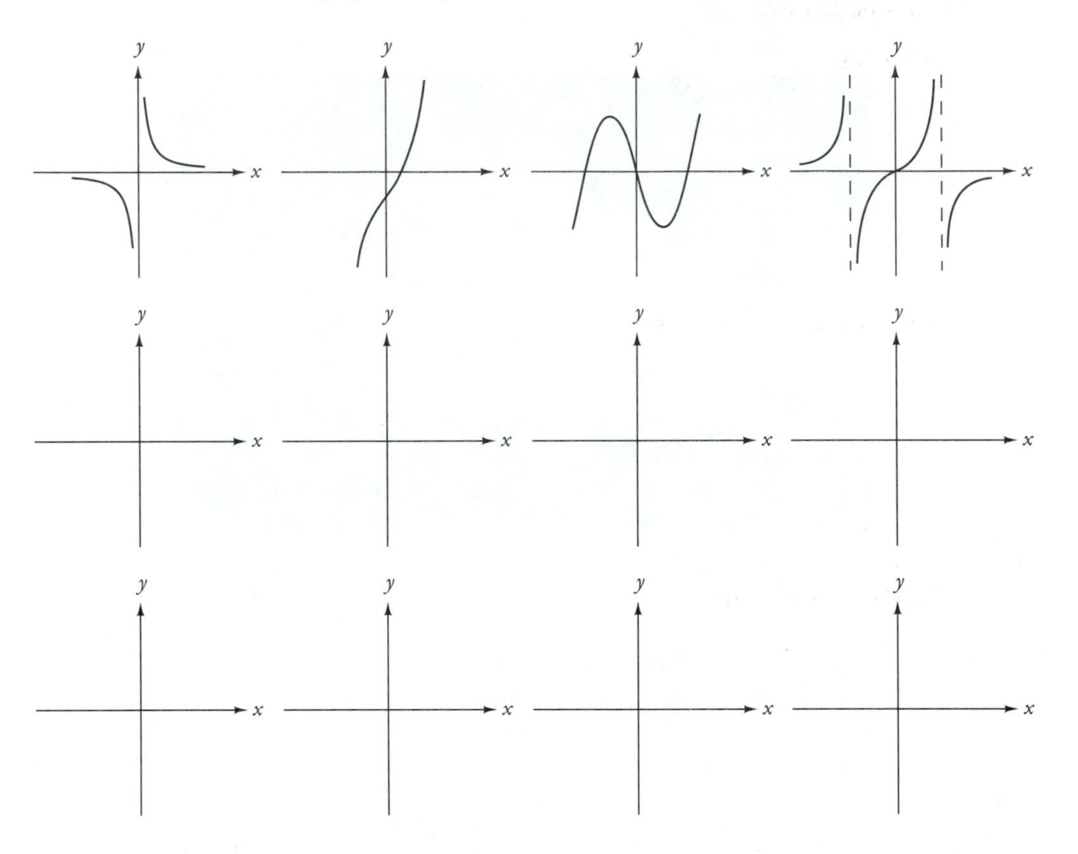

Examine the eight graphs you just drew. What do you notice?

2. Each time that you drew the graph of $f(-x)$, you did so by reflecting the graph of $f(x)$ about one of the axes. Each time that you drew the graph of $-f(x)$, you reflected the graph of $f(x)$ about the other axis.
   (a) The graph of $f(-x)$ is the reflection of the graph of $f(x)$ in the ____ axis.
   (b) The graph of $-f(x)$ is the reflection of the graph of $f(x)$ in the ____ axis.

3. Here are the formulas for the four original functions. For each one, write the formula for $f(-x)$ and the formula for $-f(x)$. Can you, for each function, make the two expressions identical using the rules of algebra? For one of the four functions, that will not be possible.

   (a) $f(x) = \dfrac{1}{x^3}$

   (b) $f(x) = x^3 + 2x - 3$

   (c) $f(x) = x^3 - 4x$

   (d) $f(x) = \dfrac{x}{4 - x^2}$

4. In your own words, explain why you could have expected such results simply by looking at the four given graphs.

The formula for a function is a giveaway for this type of symmetry, too. Whenever $f(-x)$ is *algebraically equivalent* to $-f(x)$ (that is, when we can make them the same using rules of algebra), the graph of $f(x)$ will be symmetric about the origin.

5. Test the following algebraically to predict whether or not their graphs will show origin symmetry. Check your results by looking at their graphs.
   (a) $7x - 3x^3$

   (b) $x^5 - 25x^3 + 100x$

   (c) $x^5 - 25x^3 + 100$

   (d) $\dfrac{3}{x-3}$

   (e) $\dfrac{x}{x^2-4}$

   (f) $\dfrac{x^2-4}{x}$

6. Does a graph with origin symmetry necessarily pass through the origin? Explain.

7. Suppose the graph has origin symmetry *and* a y-intercept. Does it pass through the origin? Explain.

3.5: THE RULES OF THE GAME—Testing for Symmetry

This project is a follow-up to Lab 3 and depends upon 3.3 and 3.4

Let's summarize what we learned in the two symmetry investigations. If we know the formula for a function, we can test for y-axis or origin symmetry by evaluating $f(-x)$.

- If $f(-x)$ is algebraically equivalent to $f(x)$, the graph has symmetry about the y-axis (also known as "even" symmetry).
- If $f(-x)$ is the opposite of $f(x)$ (that is, if $f(-x)$ and $-f(x)$ are algebraically equivalent), the graph has symmetry about the origin ("odd" symmetry).
- If $f(-x)$ is something else, the graph has neither type of symmetry. (The graph might possibly be symmetric about a different line or a different point, but we do not call it "even" or "odd.")

Apply the symmetry test to the following functions. Predict whether their graphs will have y-axis symmetry, origin symmetry, or neither. Then check your results by looking at the graphs. In every case, you should start by computing $f(-x)$.

1. $r^4 + \pi$

2. $r^2 + \pi r$

3. $\sqrt{t}$

4. $\sqrt[3]{t}$

5. $2 - s + s^3$

6. $\sqrt{x^4 - 9}$

7. $\dfrac{x}{x-3}$

8. $\dfrac{s}{3-s^2}$

9. $\dfrac{z^2-9}{z^2+2}$

10. $\dfrac{\sqrt[3]{t}}{t^3}$

11. $\dfrac{z^3-9}{z^2+2}$

12. $\dfrac{s^2+3}{s}$

13. $\dfrac{x^2-4}{x^2-2}$

14. $\dfrac{x^3}{x^2-2}$

3.6: ABSOLUTELY!—Absolute Value in Functions

This project is a follow-up to Lab 3

By now you should be getting the idea that success in graphing depends in large part upon learning a few basic shapes and then being able to recognize patterns. This assignment will teach you how to toss absolute values into the mix.

$|f(x)|$: THE ABSOLUTE VALUE OF A FUNCTION

The absolute value of a quantity is always nonnegative. The absolute value of an entire function, therefore, will have to be nonnegative as well. Keep this in mind as you investigate these functions.

1. Have your grapher draw both $f(x) = x^2 - 4$ and $g(x) = |x^2 - 4|$ on the interval $-3 \leq x \leq 3$. Sketch what you see, using two different colors and showing clearly where the graphs coincide and where they differ.

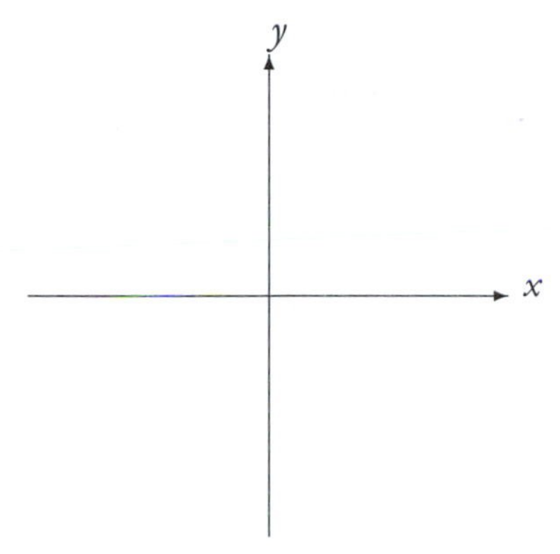

   Explain what the absolute value did to the graph of the second function. Why did some portions of the graph change, while others remained the same?

2. In the same manner, graph $f(x) = 4 - x^2$ and $g(x) = |4 - x^2|$. Compare your results with what you saw in the first set of graphs. Are you surprised? Explain why two of the graphs are identical.

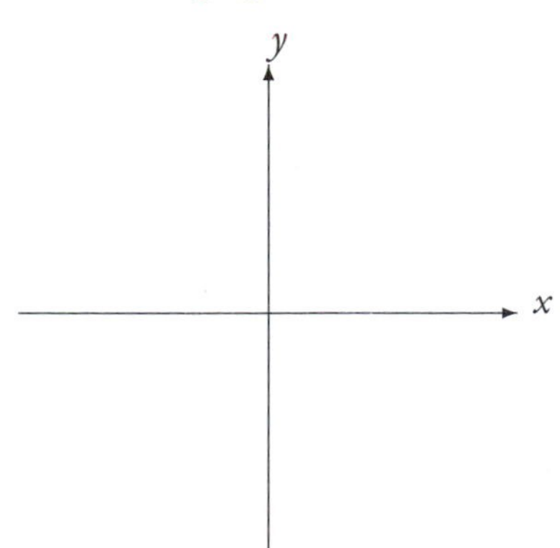

3. For each function given, sketch a quick graph. Then, write the formula for $g(x) = |f(x)|$ and use your sketch to make a graph of $g(x)$. Check your work with your grapher.

(a) $f(x) = x - 2$

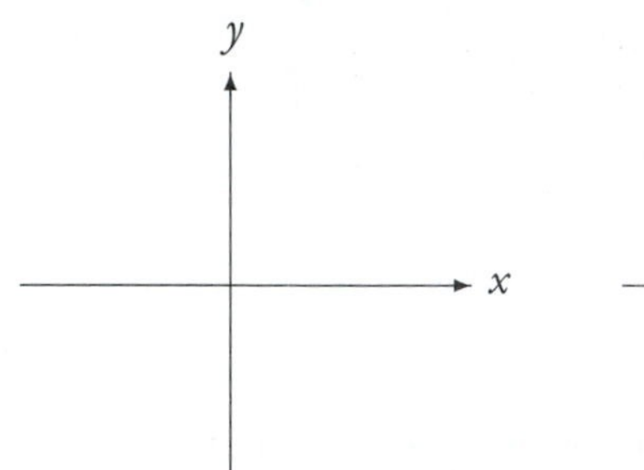

(b) $f(x) = \frac{1}{x}$

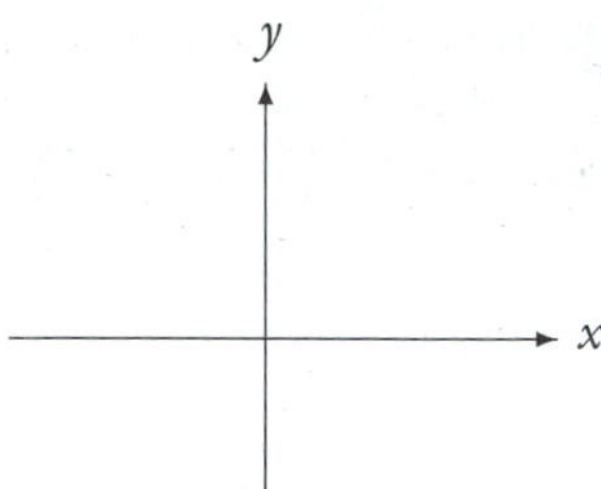

(c) $f(x) = \sin(x)$

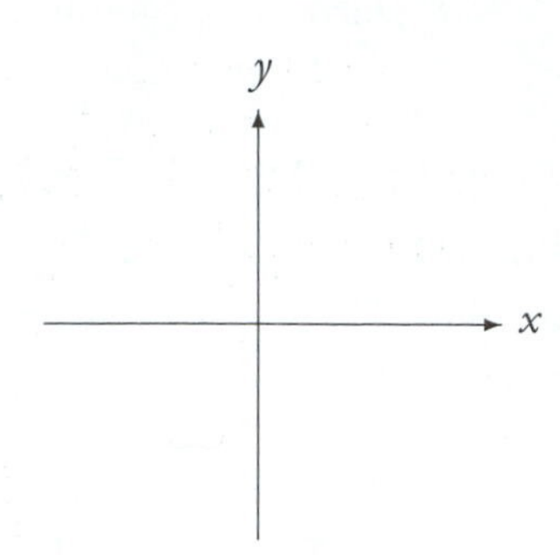

4. If the graph of some function $KIWI(w)$ looks like the graph below, sketch the graph of $|KIWI(w)|$.

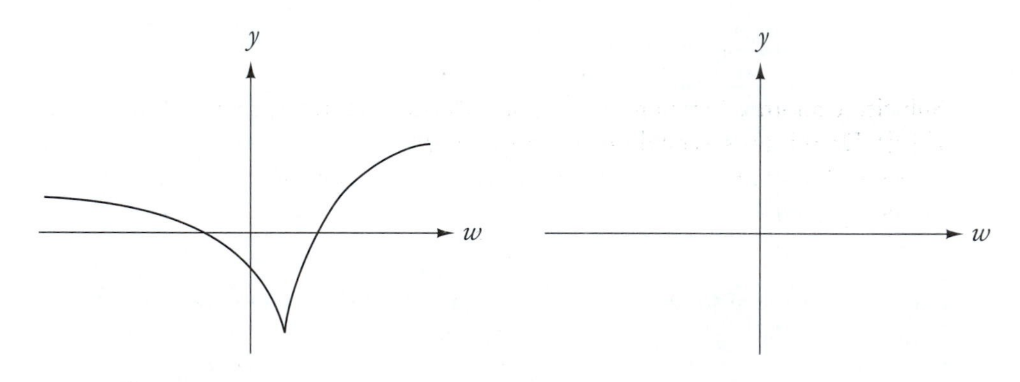

5. Write a sentence or two summarizing what you have learned about taking the absolute value of an entire function. (In the next section you will learn what happens to a function when the absolute value is applied first.)

$f(|x|)$: THE FUNCTION OF AN ABSOLUTE VALUE

Something different happens if we apply the absolute value to the *independent* variable. The expression $f(|x|)$ implies that, regardless of whether the input variable is negative or positive, the function will use only the *magnitude* (a nonnegative quantity) of that variable. For example, if x happens to be -5, the value to be computed will be $f(|-5|)$, or $f(5)$.

6. Let's see how this affects a graph. Draw the graph of $f(x) = x^2 - 2x$ on the interval $-4 \leq x \leq 4$. Write the formula for $g(x) = f(|x|)$, and overlay the graph of $g(x)$. Sketch what you see, using two colors and showing clearly where the graphs coincide and where they differ.

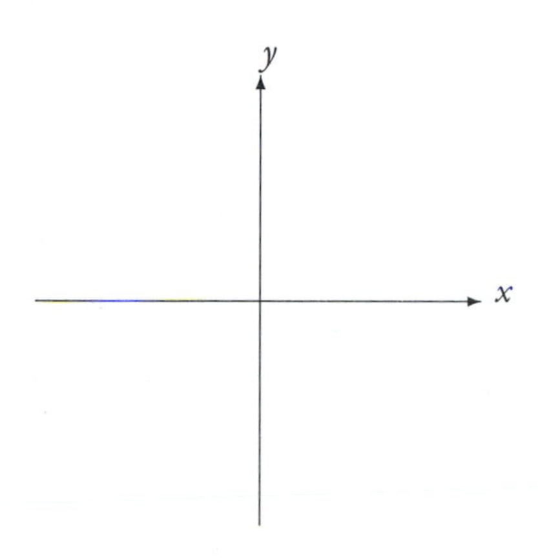

7. Substitute another function for $f(x)$ and view its graph together with the graph of $f(|x|)$. Do this for several different choices of $f(x)$ until you feel certain of the effect of using absolute value in this way. Describe what happened, and explain why the graph of $f(|x|)$ looks the way it does.

8. You should observe that some of the outputs are still negative (that is, below the x-axis). Explain why, even though we're dealing with absolute values, we can still end up with negative values for the *function.* Why didn't that occur in the last section, when you took the absolute value of an entire function?

9. Here are some practice problems. For each function given, sketch a quick graph. Then, write the formula for $g(x) = f(|x|)$, and use your drawing to make a graph of $g(x)$. Use the grapher to check your work.

(a) $f(x) = x - 2$

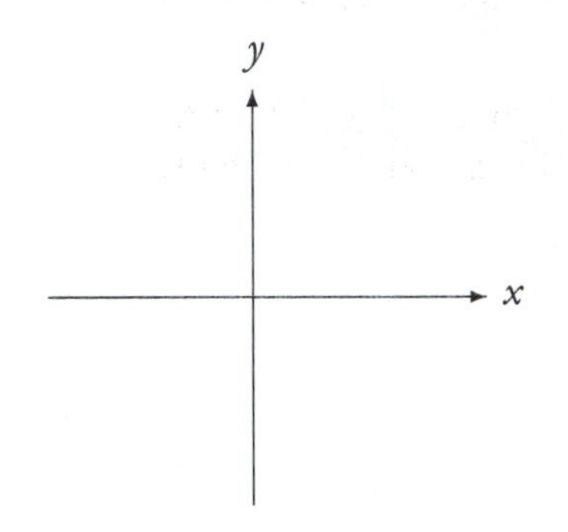

(b) $f(x) = \sqrt{x + 1}$

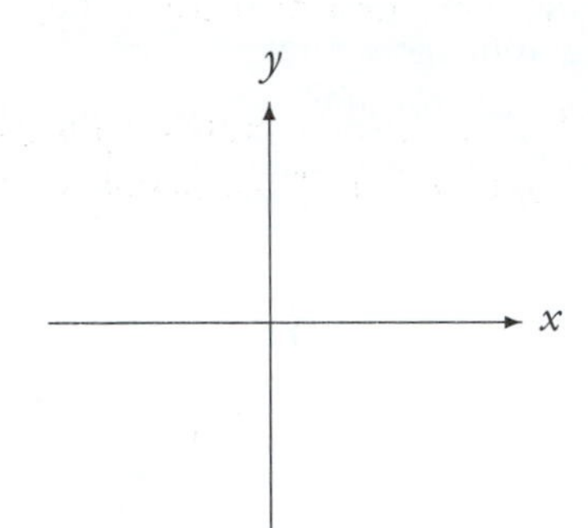

(c) $f(x) = \sin(x)$

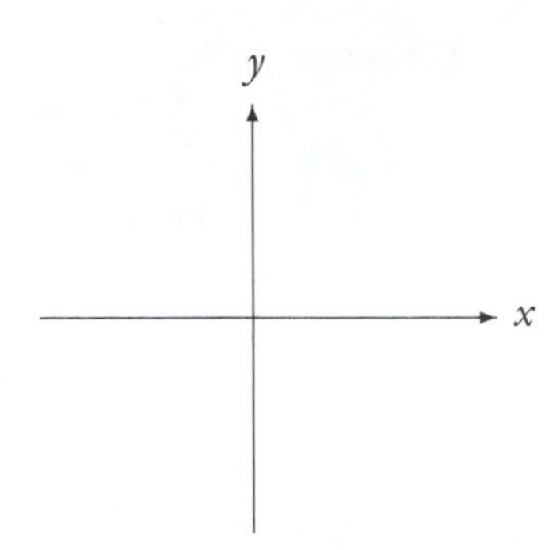

10. Using the graph of $KIWI(w)$, sketch the graph of $KIWI(|w|)$.

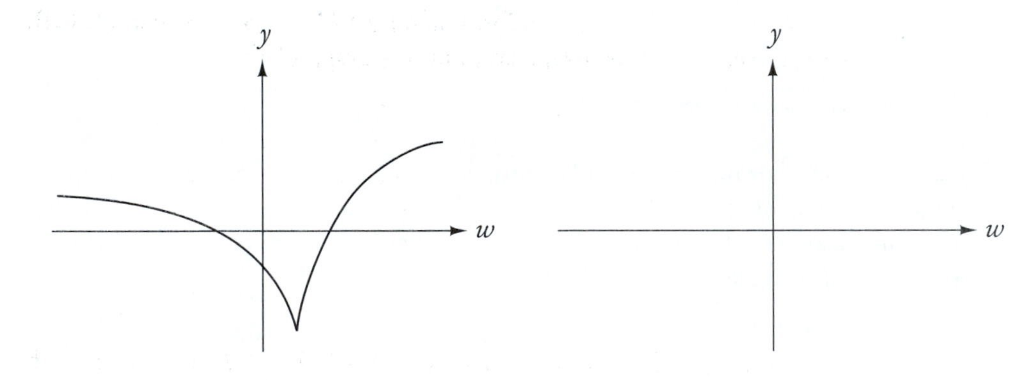

11. Explain what you did, in each case, to the graph of $f(x)$ to produce the graph of $f(|x|)$.

12. What type of symmetry do you observe in any $f(|x|)$ graph?

3.7: BEAUTIFUL COMPOSITIONS—Linear Transformations

What happens to the graph of any function $f(x)$ when you compose $f(x)$ with a linear function or compose a linear function with $f(x)$? That's what you will discover in this investigation.

1. Start with two simple functions, a quadratic function $f(x) = x^2$ and a linear function $g(x) = x + 3$.
   (a) First, compose f with g. Write a formula for $f \circ g(x)$, that is, $f(g(x))$. Now graph $f(g(x))$ and $f(x)$ together. What happened to the graph of f when it was composed with g?

   (b) Next, compose g with f. Write a formula for $g \circ f(x)$, also known as $g(f(x))$. What is the effect of this composition on the graph of f?

   (c) What are the domain and range of f? Of $f \circ g$? Of $g \circ f$? How does each composition affect the domain and the range?

   (d) If you were to compose any other quadratic function with any other linear function, would the result necessarily be a quadratic? Suppose you composed the functions in the opposite order. Do you think the result would still be quadratic?

2. Repeat the steps in (1) using $f(x) = \sqrt{x}$ and the same linear function. When you get to part (d), ask yourself whether the composed functions will necessarily be members of the square root family of functions.

3. Now it's time to generalize. Let $f(x)$ be any function and $g(x) = x + c$ any linear function with slope 1. Answer the following, supporting your answers with several examples. In your examples, please consider both positive and negative values for c.
   (a) How does the composition of f with g affect the graph? The domain? The range?

   (b) How does the composition of g with f affect the graph? The domain? The range?

4. All of the linear functions so far have had a slope of 1. Now use $f(x) = \sin(x)$ and $g(x) = 3x$.
   (a) Write a formula for $f \circ g(x)$ and one for $g \circ f(x)$. Now, graph $f(x)$ together with the two compositions.

   (b) What happened to the graph of f when you composed f with g? What happend when you composed g with f?

   (c) Which of these compositions affected the range of f?

5. Suppose $f(x) = \sqrt{x-1}$. This time, let $g(x)$ be $\frac{1}{2}x$.
   (a) Write a formula for $f \circ g(x)$ and one for $g \circ f(x)$. Now, graph $f(x)$ together with the two compositions.

   (b) What happened to the graph of f when you composed f with g? What happened when you composed g with f?

   (c) Which of these compositions affected the domain of f?

6. It's time for the full generalization. Let $f(x)$ be any function and $g(x) = mx + b$ be any linear function. Answer the following, providing several examples as supporting evidence.
   (a) What will happen to the graph of $f(x)$ when you compose f with g?

   (b) What will happen to the graph of $f(x)$ when you compose g with f?

Lab 4: Energy

You Light Up My Life

PREPARATION

It is a hot day in the middle of August. You are typing away on your computer (things are going great) and all of a sudden . . . the power drops! Hours of work are down the drain. (You never save as you type!)

During times of prolonged extreme temperatures, electric companies are not always able to keep up with demand. Blackouts and brownouts occur, causing major and minor power disruption. In order to provide better service to their customers, electric companies routinely investigate the relationship between energy usage and temperature.

Utility companies also supply data to customers to help them recognize patterns of electrical usage in their homes or businesses. The two charts below were included with the electric bills of two Massachusetts homeowners.

- One of the houses has electric heat and consequently uses much more electricity in the winter than the other, which has gas heat. Which energy profile represents the home with electric heat? How did you decide?
- One of the houses has an electric air conditioner and therefore uses its greatest amount of energy in August. Which profile represents that house? How do you interpret the peaks in its energy profile?
- Which of the two houses belongs to a professor who closes the house for most of January while he is in Cleveland? Support your answer with a statement about energy use in that home.

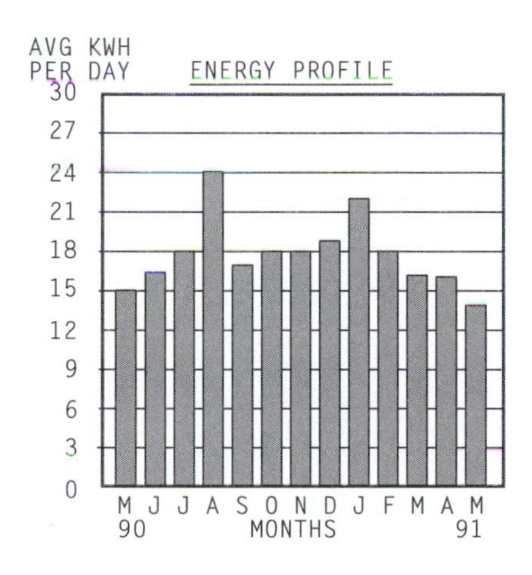

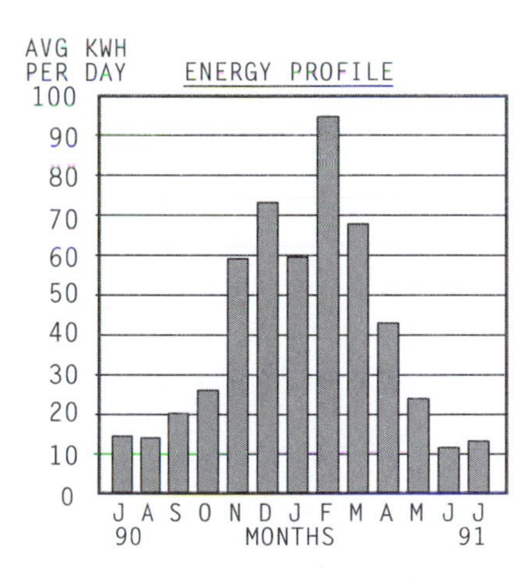

These two profiles illustrate the fact that temperature affects energy consumption. Extremes of temperature in either direction cause increased demands on electric utilities, as we saw in January, 1994, when the Northeast and the Midwest experienced a mini Ice Age.

Now think about how temperature influences energy consumption in general, that is, for the entire region served by a power company. Make a rough sketch of a graph illustrating how the overall energy usage in such a region depends on *temperature* for temperatures between 55 degrees and 100 degrees Fahrenheit (that is, for typical summer temperatures). Because it depends on temperature, *energy consumption*, measured in kilowatt-hours, will be the dependent variable. For your sketch, you can use a vertical scale of "a little," "a medium amount," and "a lot." What is the independent variable?

How did you decide whether the shape of your graph should be a line or a curve? In plain English, and in the context of electric energy, what's the difference between a straight-line graph and a curve?

Now sketch a second graph, with the same independent and dependent variables, but this time illustrating the relationship between temperature and energy consumption using temperatures typical of winter, perhaps from -10 degrees to 50 degrees Fahrenheit. Does your graph show an increasing function or a decreasing function?

Finally, piece together your graphs to get a single graph that summarizes the relationship between temperature and energy consumption. Use the same rough scale on the vertical axis that you used for the first two graphs. Bring this graph to the lab, along with your ideas about curved graphs versus straight graphs, so that you can compare results with your partners.

THE ENERGY LAB

Electric utilities keep records of the amount of energy consumed every month and the average temperature for that month. From actual data supplied by Massachusetts Electric Company, we have constructed two mathematical models predicting monthly energy consumption based on average monthly temperature.

You won't see the numbers themselves, but you'll work with functions that were developed from the data using statistical techniques. Both models are reasonably good approximations of the data.

Model 1: The Piecewise Linear Model

If you decided that the model relating electrical energy consumption to temperature was a V-shaped graph of two lines pieced together, one representing the relationship during the warm weather and the other for the cooler weather, the model partially specified below might be close to the shape of the graph you drew in preparation for this lab.

For the cooler temperatures, use

$$p_1(x) = 1588 - 8.7x$$

and for the warmer temperatures, use

$$p_1(x) = 390 + 11.4x$$

where x is the average monthly temperature, measured in degrees Fahrenheit, and $p_1(x)$ is the resulting energy consumption, measured in millions of kilowatt hours.

Note: there are differences between the graphs you drew in the preparation section and the two graphs you will study in this lab. One important difference in the models presented here is that the only temperatures used are average monthly temperatures (the average, over the entire month, of the daytime and nighttime temperatures), so you probably won't look at temperatures as extreme as those you might have considered earlier. Another difference is that, in the preparation section, you had no way to measure the energy consumption; you could indicate on your graph only that the amount for one temperature was higher or lower than the amount for a different temperature. Nevertheless, the general shape of the graph you drew should be comparable to one of the models presented here.

Graph the two lines that make up $p_1(x)$. At what temperature do these lines meet? Complete the algebraic definition of the model by filling in the intervals below.

$$p_1(x) = \begin{cases} 1588 - 8.7x & \text{for } x \text{ \_\_\_\_\_\_\_\_\_\_\_\_} \\ 390 + 11.4x & \text{for } x \text{ \_\_\_\_\_\_\_\_\_\_\_\_} \end{cases}$$

In other words, specify the temperature interval on which the each formula applies.

In terms of what happens in the real world, explain why the coefficient for x is negative in one portion of Model 1 and positive in the other portion. (Think about what the slopes of the lines signify in the context of temperature and energy consumption: "As temperature rises, ...")

Model 2: The Parabolic Model

If you decided that the relationship between energy consumption and temperature could best be represented by a curve, your sketch may have more closely resembled the graph of this model:

$$p_2(x) = 1953 - 28.7x + 0.251x^2$$

Comparing the Two Models

Graph $p_1(x)$ and $p_2(x)$ on the same viewing screen. Many graphers allow you to define a function in piecewise fashion. If yours has this capability, find out how to display only the segment of each line that belongs to the model $p_1(x)$. Otherwise, you will have to keep track of which line to use for which temperatures.

Suppose the average monthly temperature for July was 75° F. Estimate energy usage with Model 1 and with Model 2. Find this information first from your graph. Then, find the answers using algebra.

For what temperatures do the two models predict the same energy consumption? Solve this problem using your graph. Now write down the algebraic equivalent of the question and use algebra to check the solution. The *algebraic equivalent of the question* is an equation. To solve it, you will need your calculator and the quadratic formula.

For what temperatures does Model 1 predict higher energy consumption than does Model 2? Solve this problem using your graph. Now write down the algebraic equivalent of this question. In this case, the algebraic equivalent of the question is an inequality. (How will you handle the fact that $p_1(x)$ consists of two pieces?) To solve the inequality, you should not do any additional algebra. The graph will show you the intervals that form the solution.

Restrict your viewing screen so that temperatures range from 16° F to 77° F. (These are the minimum and maximum average monthly temperatures recorded over a ten-year period by the Massachusetts Electric Company.) How different from one another do the two models appear over this temperature interval? For which temperatures would you say the two models exhibit the most discrepancy in predicting energy consumption?

The Difference Function

At 75° F, what was the discrepancy in energy usage between Models 1 and 2? You have just computed $p_1(75) - p_2(75)$, the **difference** between the two models for $x = 75$. To compare Model 1 with Model 2 for other temperatures, you can form the function

$$d(x) = p_1(x) - p_2(x)$$

(Aren't you glad you don't have to do the subtraction for each temperature individually?)

Write down the algebraic expression for $d(x)$. (Be careful. Since $p_1(x)$ is a piecewise function, $d(x)$ will also need to be defined as a piecewise function.) To view the graph of $d(x)$, you might be able to avoid typing in all those symbols by instructing your grapher to define it in terms of the other two functions.

Put $d(x)$ on the screen along with $p_1(x)$ and $p_2(x)$.

Use the graph of $d(x)$ to find the average monthly temperatures for which the two models predict the same energy consumption. Compare the answer you obtained using $d(x)$ with your answer computed using the graphs of the two models.

How can you use the graph of $d(x)$ to determine the temperature interval(s) for which Model 1 predicts higher energy usage than does Model 2? Verify that this approach yields the same solution that you obtained before.

The function $d(x)$ combines much of the information from Models 1 and 2 and gives us a handy tool for comparing them. Some information, however, is lost. What information were we able to obtain from Models 1 and 2 that we can't get from $d(x)$?

THE LAB REPORT

This lab presented two mathematical models for predicting energy usage.

- Explain how each model represents the reality. What's the significance of the shape? What does each one tell you about patterns of energy usage as affected by temperature?
- Discuss their similarities and differences. Be sure to explain, in terms of energy usage, the difference in meaning between the straight line and the curve.
- Show how you used the graphs to estimate the solutions to an equation and an inequality, and explain the connection between those graphical solutions and the solutions you obtained using algebra.
- Explain how you used the difference function $d(x)$ to compare the two models.
- Illustrate your discussion with clearly labeled graphs. (A single graph showing all three functions might be the best way to present your findings.)
- Include, if you wish, comments on the graph you drew as preparation.

Section 4—Projects and Explorations

4.1: ONE OF THESE THINGS IS NOT LIKE THE OTHER—Equations and Inequalities

When you studied algebra, you learned many equation-solving techniques, such as factoring and using the quadratic formula. You learned to solve simple linear inequalities and perhaps some more complicated ones involving quadratic expressions. You might have gotten the idea that most equations and inequalities can be solved algebraically, simply because all of the ones you happened to see had algebraic solutions. The bitter truth is that most polynomials cannot easily be factored and that the solutions to many equations cannot be readily determined using the methods of algebra. Thanks to graphing technology, though, we can easily find excellent *decimal approximations* to the actual solutions of any equation or inequality we're likely to encounter.

In this investigation, you will learn to use your grapher to solve equations and inequalities, many of which you would be unable to solve using algebra alone. You will also learn to translate questions about functions into equations or inequalities that you can solve with your grapher.

Using a Graph to Solve an Equation

1. Here's an equation to solve: $x^3 - 3x = x^2 + 7$. Although it doesn't look complicated, none of the algebraic tricks you might have learned will help you. Try this instead: graph the two functions $f(x) = x^3 - 3x$ (the left side of the equation) and $g(x) = x^2 + 7$ (the right side), and locate their intersection. The x-value of the intersection point is the x that solves the equation. You won't find the *exact* solution, but you can zoom in on the point until the result is as close to the actual solution as you wish it to be.

2. Check your result by substituting it into the equation. You won't obtain exact equality, because your number is only a decimal approximation to the actual solution, which is an irrational number. If you don't think the approximation is good enough, zoom in more closely.

3. Let's be more specific about accuracy. Suppose we want an answer accurate to four decimal places. Zoom in repeatedly on the intersection until the digits in the first four decimal places of the x-value do not change as the cursor moves across the point of intersection. Write down the x-value, using only four decimal places. In this particular case, the number in the fifth decimal place is definitely more than 5, so you should round up. The result should be 2.8829. Is that what you got?

4. When we say that the value $x = 2.8829$ is accurate to four decimal places, we mean that the true solution lies in the interval $2.88285 \leq x < 2.88295$. Convince yourself that 2.8829 is better than 2.8828 or 2.8830 by substituting all three values into the equation. None gives a perfect match. Which one is best?

5. Here's another graphical method that you might prefer. Rewrite the equation as $x^3 - 3x - x^2 - 7 = 0$. Then make a function out of the left side and see where its graph crosses the x-axis. You should, of course, obtain the same result, because this equation is equivalent to the original one. (Many calculators will approximate the roots of a function. See pages 241, 271, and 300.)

6. For what values of x does the function $f(x) = x^2$ give the same outputs as the function $g(x) = 2^x$? We can turn this question into an equation, $x^2 = 2^x$, and solve for x. Feel free to break your head over this if you want, but there is no algebraic method of solving it. You might find two solutions using trial and error, but there is a third, irrational, solution for which algebra gives us no help. Use graphing techniques to find it, and give the value correct to three decimal places. Find the other two solutions (which are *exact*) by any method you wish.

7. Check all three solutions by substitution. How can you tell that two are exact and one is approximate?

Using a Graph to Solve an Inequality

8. With a grapher, solving inequalities isn't much more difficult than solving equations. Suppose we need the solution to $x^3 - 3x > x^2 + 7$. Once again, we'll create two functions: $f(x) = x^3 - 3x$ and $g(x) = x^2 + 7$. We already know that they intersect only at $x = 2.8829$. All we need to know now is whether $f(x)$ is above $g(x)$ to the right of that point or to the left. If right, then our solution is $x > 2.8829$; if left, then $x < 2.8829$. Which is it?

9. With more than one point of intersection, there are more intervals to consider. Use your grapher to solve the inequality $x^2 < 2^x$. Be careful; the solution consists of two separate intervals. Tell how you used the graph to determine the answer.

10. Suppose, as a mathematician, you are interested in comparing two functions: a linear function $f(x) = 4x + 1$; and a polynomial function $g(x) = x^3 - 3x^2 + 1$. For what values of x does the linear function produce smaller values than the polynomial function? Translate this question into an algebraic inequality and solve it with your grapher. Do you think that your solution is approximate or exact?

11. The preceding inequality does have an exact solution. Find it by changing the inequality into an equation. Use algebra to solve the equation (factoring will help), and recognize the solutions as the x-coordinates of all of the points of intersection of f and g. These values of x are the *only* values for which the graphs intersect. Now deal with the inequality: on which intervals is the graph of $f(x)$ below the graph of $g(x)$? Compare this solution to the one you obtained graphically.

A Little Practice

12. Find the solution sets to the following equations and inequalities. You choose the method (graphing or algebra), and state whether your solution is exact or approximate. Find all approximate solutions correct to three decimal places.

    (a) $|3x - 5| = x + 7$

    (b) $|3x - 5| < x + 7$

    (c) $\sin(x) = (x - 1)^2$

    (d) $\sin(x) > (x - 1)^2$

    (e) $x^2 + 5 < 0$

    (f) $2x^2 + 3x - 4 < 4x - 1 < x^2 + x$

The last one's tricky. The solution consists of those intervals on which the first graph is below the second, and, at the same time, the second is below the third.

4.2: HOW FITTING!—The Least Squares Line

This project is a follow-up to Lab 4

The Energy lab presented two models for predicting energy consumption from the average monthly temperature, but it did not tell you how either model was derived. In this project, you will see the mathematical ideas behind the piecewise linear model and learn to use a calculator to produce your own best-fitting line for a set of data.

You probably noticed that, in the Energy lab, you did not see any actual data. At that point, tables of data would probably have distracted you from the goals of the lab, so we did not include them. To create the models, we used data provided by the Massachusetts Electric Company, which has an entire department devoted to the collection and processing of information about energy usage.

This table shows a sample of the kind of data the company collects. For simplicity, we have rounded all values to the nearest integer.

Table A

Average monthly temperature (degrees Fahrenheit)	Energy consumption (millions of kilowatt-hours)
63	1048
64	1110
67	1134
72	1155
76	1312

1. Plot these points on graph paper, using x as the variable for the first column and y for the second column. Draw a straight line that you think best represents and summarizes the data you plotted. (You won't be able to force the line to pass through all the points.) Determine the linear function for the line you drew.

2. Now you'll do a little statistical analysis on the data. Begin by completing Table B. "Predicted y" means the y-value that your equation produces for the given x-value. "Residual error" is the difference between the value we actually have and the one that the linear function gives. Subtract the third column from the second to find the residual. For the last column, square the values in the fourth column.

Table B

x	y	Predicted y	Residual error (y − predicted y)	(Residual error)^2 (y − predicted y)^2
63	1048			
64	1110			
67	1134			
72	1155			
76	1312			

3. Finally, add up the last column and write the total.

How are we to decide whether or not this line is a good fit for the data? There are many ways of selecting a line to represent a set of data and, if you compare your line with those of other students, you'll probably find that no two are exactly alike. To avoid this ambiguity, statisticians frequently follow what is called the **least squares** criterion for selecting a single line to represent a data set.

When you calculated the residual errors for your line, you probably had some negative values and some positive values. If we added those numbers, they could conceivably sum to a value close to zero, implying hardly any discrepancy between the data and the linear model; yet the points would deviate from the line in both directions. To sidestep the issue of signs, statisticians use the *square* of each error, because the main concern is how far from the predicted value each point is, rather than whether it's above or below.

The least squares criterion says to select the line for which the total of the squared residual errors is as small as possible. (This total is also known as the sum of the squared errors, or SSE.)

4. What is the SSE for your linear function?

Now let's see how well your line compares with what a statistician would consider the line of best fit. Use your calculator or statistical software to compute the least squares (also called the linear regression) equation. See the guides for the T1-82/83 (page 225), TI-85/86 (page 253), and TI-92 (page 283) graphing calculators, or consult the instruction book for your calculator. Many graphing calculators, given the information in Table A, will generate the slope and the y-intercept of the least squares line.

5. Write the equation of the least squares line. Round the slope value to three decimal places and the intercept to one decimal place.

6. Complete Table C, using the least squares line for the "predicted y."

Table C

x	y	Predicted y	Residual error (y − predicted y)	(Residual error)^2 (y − predicted y)^2
63	1048			
64	1110			
67	1134			
72	1155			
76	1312			

7. Compare the SSE for the least squares line to the SSE for the linear function you wrote in (1). Why do you think that statisticians would prefer the line with the smaller SSE?

8. Use a grapher to view the data, the least squares line, and your original linear function. Now that you see everything, do you agree that the least squares line does a better job of capturing the overall trend of the data?

Table D shows the actual data used for the warm-weather segment of the piecewise linear model from the Energy lab. The numbers represent information from the the months of June through September of 1986–1989.

Table D

Average monthly temperature (degrees Fahrenheit)	Energy consumption (millions of kilowatt-hours)
63.23	1048.40
68.48	1055.94
71.81	1088.53
64.83	1081.63
62.77	1121.68
64.39	1110.17
72.03	1155.41
66.87	1134.52
63.53	1166.57
71.29	1215.95
75.61	1312.37
68.70	1198.38
64.13	1193.52
70.42	1206.26
72.58	1267.14
68.30	1254.12

9. Use a calculator or a computer program to determine the least squares line for these data.

10. Compare that equation to the warm-weather portion of the piecewise linear model in the Energy lab, $390 + 11.4x$. Now you know where it came from!

4.3: SKELETON KEYS

Plan ahead, or this project will take too long! Unless your instructor is able to link calculators and transfer the data, work together. Each group member does a different equation in (1). Share the results.

Physical anthropologists can determine much about a deceased person from the bones in his or her skeletal remains. From the femur (thigh bone) or the ulna (the inner bone of the forearm), for example, a scientist can estimate the person's height.

In this project, you will play the role of a physical anthropologist and solve the mystery of the bones. The equations you find will be very similar to those proposed by Dr. Mildred Trotter, who served as a special consultant to the U.S. Government during World War II. Her task was to identify dead soldiers from their skeletal remains. Forensic scientists and law enforcement agencies still use some of Dr. Trotter's formulas for estimating a person's height from the lengths of his or her bones.

The following data come from the Forensic Data Bank at the University of Tennessee. There were 29 female and 31 male skeletons represented.

Females				Males			
No.	Height (cm)	Ulna (mm)	Femur (mm)	No.	Height (cm)	Ulna (mm)	Femur (mm)
1	168	258	448	1	177	279	488
2	161	227	413	2	175	272	464
3	158	237	432	3	181	290	487
4	173	265	473	4	171	268	454
5	163	236	428	5	175	271	470
6	168	244	441	6	183	281	505
7	168	246	448	7	167	264	447
8	165	246	435	8	173	276	463
9	173	266	483	9	180	288	485
10	163	250	450	10	180	278	494
11	158	232	414	11	171	260	448
12	163	245	443	12	175	272	456
13	170	253	440	13	171	258	449
14	159	233	419	14	168	262	444
15	170	253	449	15	177	266	483
16	165	252	443	16	180	281	490
17	165	249	451	17	178	278	477
18	165	240	448	18	178	271	487
19	168	248	450	19	181	282	488
20	163	236	435	20	170	266	459
21	165	250	448	21	173	261	460
22	163	244	434	22	173	261	460
23	165	248	452	23	183	285	502
24	177	279	488	24	178	278	480
25	171	270	461	25	177	282	482
26	162	245	421	26	166	258	442
27	180	281	482	27	180	290	505
28	180	279	486	28	175	274	470
29	175	273	484	29	171	270	461
				30	172	278	449
				31	180	281	482

1. We wish to be able to predict a person's height from the length of one of his or her bones, either the femur or the ulna. Thus, *height* will be the dependent variable and *length of femur* (or *length of ulna*) the independent variable. The table suggests four equations representing the four different relationships, two for men's heights and two for women's heights.
   (a) Using statistical software or a calculator, determine the regression equation (the least squares line) for each of the four relationships. Be sure to use the correct independent and dependent variable each time. Round the slope to three decimal places and the intercept to one decimal place. Write the four equations here, labeling each.

   (b) Use a grapher to view four plots, each one showing a different set of ordered pairs and the regression line that fits those ordered pairs. Make sure that the line appears to summarize the general trend of the plotted points.
   (c) What are the units for the slope of each line?

2. Suppose that two men have femurs that differ in length by 1 cm (10 mm). By how much should we expect the men's heights to differ? (*Hint:* use the appropriate regression line.)

3. Suppose that two women have ulnas that differ in length by 2 cm (20 mm). By how much are the women's heights likely to differ?

4. Imagine that you are called in to advise law enforcement authorities on a case where all that remains of a person is a femur bone. With nothing more to go on, you are unsure whether the person was male or female. If the length of this bone is 470 mm, how much difference would it make if you used the regression equation for predicting the height of a man and it turned out that the bone belonged to a woman?

5. For what femur length would the regression equation for predicting a male's height give the same results as the regression equation for predicting a female's height?
   (a) Show how to find the answer using algebra.

   (b) Explain how to check your algebraic result using graphs.

   (c) Is that result a reasonable femur length for an actual person? (Look at the data to see what's reasonable.)

6. If a man and a woman have femurs of the same length, which of them is likely to be taller? Justify your answer using graphs of the two regression lines, remembering to stay within reasonable bounds for femur lengths.

7. For what ulna length would the regression equation for predicting a female's height give the same result as the regression equation for predicting a male's height?
   (a) Show an algebraic solution.

   (b) Tell how to check your algebraic results using graphs.

   (c) Is that a reasonable ulna length for an actual person? How do you know?

8. If a man and a woman have ulnas of the same length, which of them is likely to be taller? This situation is more complicated than the similar question in (6). Your answer will need two parts. Justify your answer using graphs.

9. Suppose that a student wanting to predict women's stature from their femur lengths inadvertently used the men's equation instead.
   (a) Based on the data we have, women's femur lengths fell in the interval from 413 mm to 488 mm. If the student used these lengths with the wrong equation to predict stature, in what interval would the predicted heights fall?

   (b) Given your answer to (a), do you think the student would notice the error? (If you have trouble visualizing metric heights, translate them into feet and inches, recalling that one inch is approximately 2.54 cm.)

10. A hunter discovers partial skeletal remains in the woods. Your expertise is needed to solve the mystery of these bones. Among the intact bones are a skull, a left and a right ulna, and a femur (found some distance from the other bones). From measurements of the skull, you are fairly certain that the person was male. The left and right ulna measure 275 and 276 mm, respectively. The femur measures 474 mm.
    (a) Predict the stature of the dead man.

    (b) Is there any reason to think that the femur came from a different individual? Explain.

4.4: IT'S NOT EASY BEING GREEN

Polyethylene film is often used for greenhouse coverings in Japan. Manufacturers of the film would like to produce as thin a film as possible so they can produce more square feet of material for less money. The customers (greenhouse owners) want a film thick enough to resist tearing from wind and rain, but not so thick that the light that passes through is insufficient to grow healthy plants. A film more than 2 mm thick will be too opaque for use as a greenhouse covering.

Assume that the cost to the consumer can be modeled by the function

$$C_1(x) = 0.3x^2 - 1.2x + 6.2$$

and the cost to the manufacturer by the function

$$C_2(x) = 0.5x^2 + 0.4x + 4.08$$

where x represents the thickness of the film in millimeters and where the cost is measured in units of \$100,000.

1. Film thicknesses for which this problem makes sense lie in the interval from 0 to 2 mm. Graph the two cost functions, restricting the x values to the interval $(0, 2]$, that is, $0 < x \leq 2$. As the thickness x increases, what happens to the cost to the consumer? What happens to the cost to the manufacturer? (Mathematicians would apply the terms *decreasing* and *increasing* over $(0, 2]$ to the functions $C_1(x)$ and $C_2(x)$, respectively.) For what thicknesses is the cost to the producer less than the cost to the consumer?

2. Japanese business philosophy (quite different from that in the United States) aims to make a quality product that suits both the manufacturer and the consumer. The government sets a standard thickness for the film, selecting the thickness that will minimize the total cost to society. One way to assess the cost to society is to form a new function, T, by taking the sum of the cost to the consumers and the cost to the manufacturers. Write a formula for the function $T(x)$, and use its graph to determine the standard thickness that the government will require.

3. If the manufacturer cheats on the standard just a bit and produces a film that is only 0.45 mm thick, how much will the company save? How much will the consumer lose? By how much will society be cheated? (Note that the amount by which society is cheated is not the value of $T(x)$ for $x = 0.45$ mm, but rather the difference between that value and the cost to society that should have been expected.)

   Don't round off your results. Use all the accuracy your computer or calculator provides and express your answers in *dollars*, recalling that C_1, C_2, and L are measured in units of \$100,000.

Lab 5: Packages

Signed, sealed, and delivered

PREPARATION

Before meeting with your group, read through the lab sheets to get an understanding of the problem as a whole. Do the volume calculations for the test-case packages (two rectangular boxes and one cylinder). Work on the volume formulas and try to come up with functions $V(X)$ for the rectangular box and $V(r)$ for the cylinder.

At the beginning of the lab period, you should compare your numbers and formulas with those of your partners and iron out any discrepancies.

$V(X) =$ ______________________________

$V(r) =$ ______________________________

The following note appeared in the June 27, 1991, *Postal Bulletin*, a publication of the U.S. Postal Service—one more example of the primal urge to standardize!

LENGTH OF UNIFORM GARMENTS/CARRIER AND MOTOR VEHICLE SOCKS AND HOSE

The following clarifications respond to field inquiries about certain provisions of the Uniform and Work Clothes Program. A future revision of the *Employee and Labor Relations Manual* will include these revisions.

Length of Shorts, Culottes, Skirts, Jumpers

Garments should not be more than 3 inches above mid-knee. Employees should not alter the length of their garments, and vendors are not authorized to make alterations that have hems falling more than 3 inches above mid-knee.

However, since everyone is not the perfect ratio of height to girth, some alterations may be necessary. Common sense must prevail in some situations. For example: If an individual stands 6 foot, 4 inches tall with a 34-inch waist, it is likely that the hem of his/her garment will fall more than 3 inches above mid-knee even when unaltered. If left as originally manufactured, the garment should still have a reasonable appearance.

Carrier/Motor Vehicle Socks and Hose

Currently, only black knee-length socks are authorized with walking shorts. However, specific requirements for skirts and culottes do not exist. Although these apparel items are an option for female employees, the National Uniform Committee intended that employees wear either the black knee-length hose, neutral-colored hose, or a coordinated, colored sock. Bright, fluorescent hose and socks are not permitted.

New socks have been designed to complement the carrier and motor vehicle uniforms and will be available for purchase soon. The *Postal Bulletin* will announce their availability.

—*Labor Relations Dept., 6-27-91*

THE PACKAGES LAB

The U.S. Postal Service will accept as fourth-class domestic parcels objects of a variety of shapes and sizes, provided that the weight does not exceed 70 pounds and that the length added to the girth of the parcel does not exceed 108 inches. The length is defined to be the measurement of whatever is the longest side of the parcel. The girth is the distance around the parcel at its thickest remaining part. When in doubt, a postal employee will refer to pictures such as these from their manual.

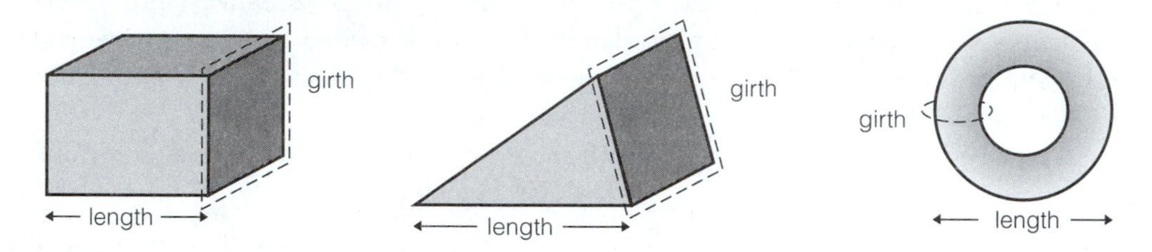

Let us suppose that we have a substance to mail, that it can be packaged in any shape whatsoever, and that we'd like to be able to ship as much of it as possible in a single package. (We will assume that the substance is sufficiently lightweight that we don't have to worry about the 70-pound restriction.)

The Rectangular Box

First, we will consider a rectangular box, and we'll suppose that the cross section of the box is square. In the Galileo lab we learned that, for a given perimeter, a square yields more area than any other rectangle, so it seems reasonable that we'd want to make the cross section of this box a square.

To check out this assumption, compute the volume for a couple of test cases. First, consider a rectangular box whose length is 44 inches, and let the cross section be a 12-by-20-inch rectangle. Verify that length plus girth equals 108 inches. What is the volume? Now, keeping the length at 44, make the cross section a 16-inch square. The volume should be greater. This, of course, doesn't prove that a square is best, but it does at least support our intuition that a "flattened" box wouldn't hold as much as a "boxy" one. (If you laid a sealed half-gallon carton of milk on its side and jumped on it, you'd have a mess, because the flattened carton can't hold as much, even though its girth hasn't changed.)

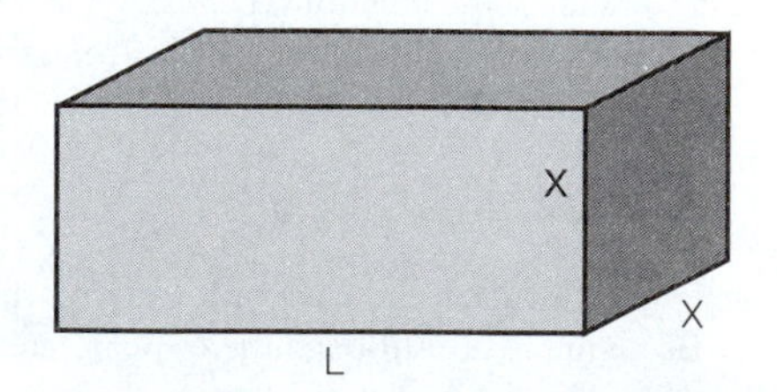

Now let both L and X vary, and write a formula for the volume of the box in terms of L and X. Then use the fact that the length plus the girth equals 108 inches (assume we'll use up the allowable 108 inches) to express the relationship between L and X.

In order to optimize V (that is, to find the greatest volume), we need to express V as a function of X alone. Use the relationship that you just wrote to assist you in writing a formula for V that has no variable other than X. As soon as your group agrees upon a formula for $V(X)$, you're ready to roll.

What type of function is $V(X)$? Use your grapher to draw its graph. Be sure to choose a viewing window large enough to see its important features: any X-intercepts, any turning points, and how it behaves beyond the intercepts. You may need to experiment with viewing windows and changes of scale before you get a meaningful picture. Is there a maximum or minimum value for $V(X)$?

You should observe that there is no maximum value for $V(X)$ because the function increases without bound as X moves away from the origin to the left. Similarly, there's no minimum value for $V(X)$. Why not? However, the graph has two turning points, one a **local minimum** and the other a **local maximum**.

Now consider only the portion of the graph that makes sense for a package. What is the smallest value of X that you would consider for the model? Remember that X represents a dimension of the package. Some values, even positive ones, are too small to be reasonable. How large an X-value would you consider? Give reasons for your answers.

Look at the portion of the graph of $V(X)$ on the X-interval you just chose. You should see a local maximum. Use the trace feature of your grapher to locate the highest point. What is the value of X at that point? (Zoom in enough to obtain X correct to three decimal places. It should be a "nice" number.) What is V? These numbers give you not only the volume of the package, but also its shape. How long should it be? What is its girth, and how did you find it?

We need to distinguish between the shape of the box and the shape of the function representing the volume of the box. The curve $V(X)$ on your screen certainly does not suggest the shape of the mailing container, yet it expresses information about the volume of the container. When X is a small positive number, what can you say about the volume of the box? As X increases, what happens to the volume? As X continues to increase beyond its optimal value, what happens to the volume? In order to visualize how the capacity and the dimensions of the box vary with the value of X, think of an object that you might ship in a box for which $X = 3$ and in one for which $X = 21$. These objects will not be at all alike!

The Cylinder

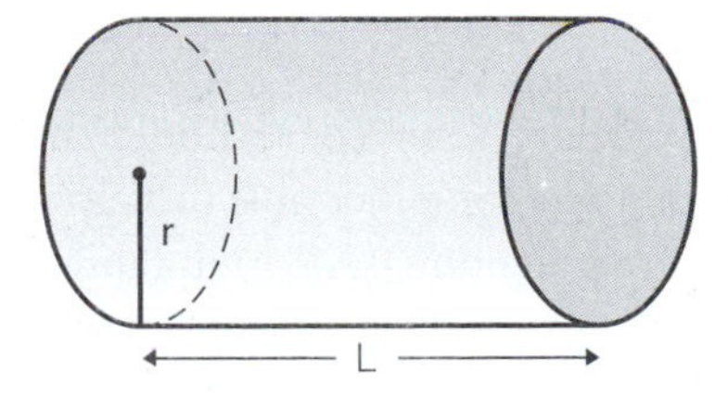

If you did *Don't Fence Me In* (Project 2.2), you saw that a circle whose circumference is equal to the perimeter of a square has a larger area than the square, so it seems reasonable that a cylindrical container satisfying postal regulations might hold more than a rectangular box. As a test, try $L = 44$ inches and see what the volume would be. (Notice that here the girth is the circumference, so you would have 64 inches for the circumference. Also, to find the volume, you need to know the radius. How can you find the radius if you know the circumference?) Compare the volume with the volumes you already calculated for the two rectangular boxes of the same length. It should be larger.

Let's try to find the best cylindrical container. As before, you will need to write a formula for the volume and then rewrite it in terms of a single variable. First write V in terms of L and r. Then use the fact that length plus girth will equal 108 inches in order to rewrite V as a function of r alone. (You'll need to write the girth in terms of r.)

What type of function is $V(r)$? Does it have a maximum or minimum value? Consider only the portion of the graph that makes sense in this context: on what interval can r represent the radius of the cylinder? Look at the graph on that interval; you should see a local maximum. Do some serious zooming to find the value of r that produces that maximum. (This value will not be a "nice" number, as the optimal X-value was, and you'll need three-decimal-place accuracy in order to make good comparisons.) Explain what the coordinates mean in terms of the package. How long should the package be? What will be its girth? Its radius?

Compare the dimensions (length and girth) of the largest rectangular box to the dimensions of the largest cylinder. Does the result surprise you, or is this what you expected?

Compare the volumes of the optimal cylindrical parcel and the optimal rectangular box. Is a circular cross section a more efficient use of girth than a square cross section? All other things being equal, which shape would you choose? Why?

The Sphere

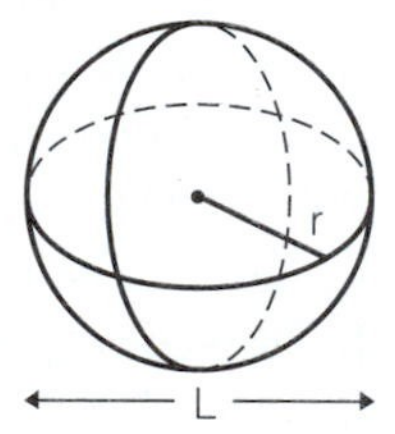

Perhaps it would be advantageous to make the package round in every direction. The Postal Service permits the mailing of spherical packages, provided they meet the weight and dimension restrictions. (Wrapping the package is, of course, a separate problem!) Perform the same analysis for a sphere, and determine the maximum volume you would be permitted to mail in a single package. What makes this problem fundamentally different from and simpler than either the box or the cylinder? Of the three shapes, which would you choose?

THE LAB REPORT

What is the best shape for a package? Your report should show how you used polynomial functions to model the volume of a container and to help answer this question. Explain the difference between the entire polynomial function (and its graph) and the portion that represents the model. Tell how you used each graph to determine the dimensions that would give the maximum volume for a particular shape, and give the maximum volume and best dimensions (length and girth) for each case. The sphere presents a situation that is different mathematically from either the box or the cylinder. Explain what's unique about the sphere. Discuss the relative merits of different shapes, under the particular restrictions of the Postal Service, and explain in everyday terms why you think one shape yields more volume than either of the other two. Illustrate your report with graphs of $V(X)$ and $V(r)$.

Section 5—Projects and Explorations

5.1: UPS AND DOWNS—Exploring Polynomial Functions

The graphs of polynomial functions show much variety, but they all share certain characteristics. In this exploration, you will learn to predict the global behavior—that is, the overall shape and direction—of the graph of a polynomial function simply by glancing at its formula. You will also learn some interesting details about these functions: their roots and turning points.

POWER FUNCTIONS: x^n

Let's begin with some very simple polynomial functions, those that have just a single term, such as x^4 and $-5x^3$. We call these **power functions;** n can be any positive integer.

1. Use your grapher to help you complete this table.

	Domain	Range	Global behavior	Turning points
x	$\mathbb{R}$	$\mathbb{R}$	down on left, up on right	0
x^2	$\mathbb{R}$	nonnegative reals	up on both sides	1
x^3				
x^4				
x^5				
x^6				
x^7				

Describe the pattern you see. Use it to predict the domain, the range, the global behavior, and the number of turning points for x^9 and for x^{20}.

2. Now modify these power functions by multiplying each by a positive or a negative constant. Which features change, and in what way? Which features remain the same? (Be sure to use negative as well as positive multipliers.)

Global Behavior of Polynomials

3. Now consider several polynomial functions of degree three, that is, functions having the form $ax^3 + bx^2 + cx + d$, where a, b, c, and d are any real-number constants, except that a cannot be 0. Compare their graphs to the graph of x^3. (Remember to try both positive and negative coefficients.) You should see that they have much in common with either x^3 or with $-x^3$. In fact, their global behavior is precisely the same as one or the other. What feature in the algebraic formula for any degree-three polynomial tells you which of the two power functions, x^3 or $-x^3$, its graph will resemble?

4. Are you ready to make a conjecture about the global behavior of a polynomial function of *any* degree? Try a few more to be sure: examine some polynomials with higher degrees and determine which power function they resemble globally.

5. Write a few sentences generalizing what you have learned about the global behavior of any polynomial function $a_n x^n + a_{n-1}x^{n-1} + \cdots + a_1 x + a_0$. (The number a_i represents the coefficient of the term x^i; all a_i are real-number constants, and $a_n \neq 0$.) You need to consider the sign of a_n and whether n is even or odd.

Roots

You probably noticed that many of the polynomial functions you graphed touched or crossed the x-axis one or more times. Each of those x-intercepts is called a **root** of the polynomial. It is often important to know about the roots of a polynomial because those are the values that cause the function itself to have the value zero. We'd like to be able to tell from the algebraic formula itself how many roots a polynomial might have.

6. The easiest way to give a polynomial roots is to create factors, one for each root. Write a formula for a polynomial function with roots at -1, 2, and 3. Sketch its graph. (Recall that you can determine its global behavior by finding its highest-degree term.)

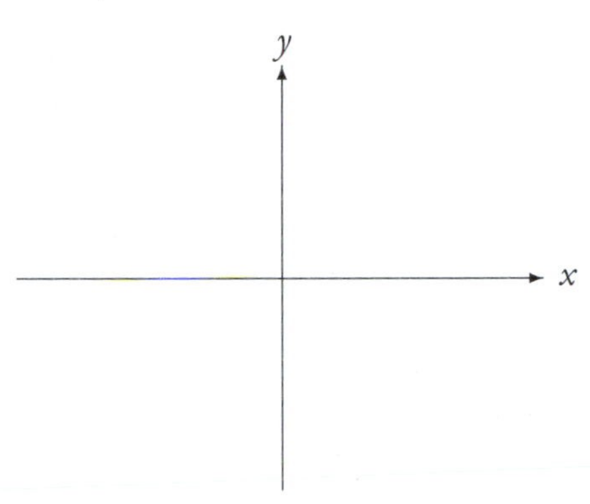

7. Was that the *only* such polynomial you could have written? Write two more polynomials that have the same roots. Let the first have degree three, and let the second have a higher degree. Sketch their graphs.

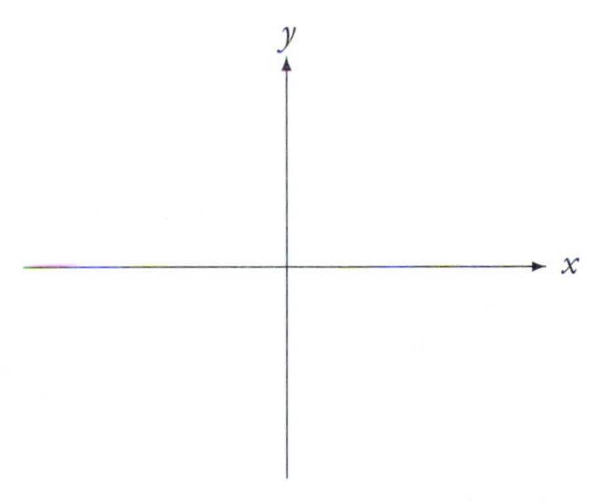

8. Write a polynomial with roots at -2, -1, 0, 1, and 2. Can you write *another* polynomial with the same roots? What is the minimum degree of either of these polynomials?

9. If the degree of a polynomial is n, what is the maximum number of roots for that polynomial?

10. Can a polynomial of degree n have fewer roots than the maximum? If you say yes, give an example. (*Hint:* look at the power functions.)

11. Can there be a polynomial with no roots? If so, give an example. If not, explain why not.

12. Can there be an odd-degree polynomial with no roots? If so, give an example. If not, explain why not.

Turning Points

A **turning point** is a point at which the graph stops rising and starts falling, or vice-versa. The functions you created in (6), (7), and (8) all had more than one turning point. Turning points are often important to locate because they give local maxima or minima, as we saw in the Packages lab.

13. How many turning points will a degree-two polynomial have?

14. How many turning points might a degree-three polynomial have? There are two answers to this question, and you might be able to figure them out by sketching some typical shapes for degree-three polynomials.

15. How many turning points might a degree-four polynomial have? This question also has two answers. Figure them out by sketching representative shapes. (*Hint:* consider x^4; consider a degree-four polynomial with four roots.)

16. How many turning points might a degree-five polynomial have? This question has *three* answers. (*Hint:* consider x^5; consider a degree-five polynomial with five roots; then picture a degree-five polynomial with fewer changes of direction.)

17. Venture a generalization about turning points. It has a lot to do with odds and evens.
    (a) Say as much as you can about the number of turning points of a polynomial of degree n, where n is odd.

    (b) Say whatever you can about the number of turning points of a polynomial of degree n, where n is even.

5.2: A HIGHER POWER—Repeated Roots

When you explored polynomial functions in 5.1, you might have noticed that, in order for a cubic polynomial to have exactly two roots, the graph needed simply to *touch* the horizontal axis at a turning point. You created a **double root** or a root of **multiplicity two** at that point. If you were to place a microscope over the spot, the bit of graph that you'd see would resemble a $\cup$ or a $\cap$. If a function with a double root is expressed in factored form, the factor for that root appears twice (that is, raised to the second power). In the Packages lab, you saw a double root at the origin for the first two volume functions. Do you remember that the first volume function had x^2 as a factor and the second one had r^2 as a factor?

1. Use your grapher to explore the effect of a double root on the graph of a polynomial function. Try a whole series of graphs. Here are some suggestions:
   (a) $y = x(x - 1)(x - 2)$
   (b) $y = x(x - 1)^2(x - 2)$
   How does graph (a) differ from graph (b)? In what ways are they the same?

2. Try keeping the factors the same, but square a different factor instead. How does the graph change?

3. Now square two or more of the factors at once and describe the resulting changes in the graph.

4. Why stop with squares? Investigate the effect of a **triple root** on the graph of a polynomial function. You will need to raise one factor to the third power. It's difficult, sometimes, to get a good picture of this from a grapher, but a triple root, under a microscope, resembles the "wiggle" in the graph of x^3 or $-x^3$. You can get some interesting pictures by starting with a simple polynomial such as (a) on the previous page and raising one factor to successively higher powers. Put a few graphs on the screen at once so that you can compare them, but keep the window narrow so that the local features aren't overwhelmed by the global stuff.

5. What do you think a root of multiplicity four would do to the graph? Find out. (You might start by comparing x^2 to x^4. Which of these is "fatter" at the bottom?)

6. How about multiplicity five?

7. Write down any conclusions you've drawn from your investigations.

5.3: MAKING A DIFFERENCE

In this project, you'll discover an intriguing numerical property of every polynomial function. There's a lot of computation to do; make this an opportunity to work with your group.

Linear Functions

1. Choose any linear function f and write its algebraic formula here. Then complete the middle column of this table by filling in function values.

x	$f(x)$	Differences
-3		
-2		
-1		
0		
1		
2		
3		

2. Now, find the difference between each pair of successive function values, and complete the third column of the table. You should always obtain m, the slope of your chosen linear function. That's no surprise! Those differences are actually average rates of change of f between successive integer values of x, and we know that linear functions have a constant rate of change.

Quadratic Functions

3. To keep the calculations simple, start with $f(x) = x^2$. Fill in the first two columns of the table, noticing that we always subtract a function value from the one that follows it: $f(n + 1) - f(n)$. The order of subtraction is important.

4. We don't expect the first differences to be equal, because x^2 isn't linear. But now calculate the second differences; you should obtain the same result—2—each time! (Remember the order of subtraction: $-3 - (-5) = 2$.)

x	x^2	First differences	Second differences
-3	9		
		-5	
-2	4		2
		-3	
-1	1		
0			
1			
2			
3			

5. Now choose another quadratic function $g(x) = ax^2 + bx + c$, letting a be something other than 1, perhaps even a negative number. Calculate function values, first differences, and second differences. Again, the second differences should be equal, though this time they will not equal 2.

x	$g(x)$	First differences	Second differences
-3			
-2			
-1			
0			
1			
2			
3			

6. Do you suppose that we could have predicted the second differences for this function? Look at the second-difference value and compare it to the value you used for the coefficient a. How are the two numbers related? Look back at x^2: what was a in that function? Was the second difference for x^2 related to the coefficient of x^2 in the same way that we see in this example?

Higher-Degree Polynomials

7. Will the pattern continue? Check it out for a cubic polynomial such as $2x^3$. First make a prediction. Then complete the table and see whether your theory holds up.

x	$2x^3$	First differences	Second differences	Third differences
-3				
-2				
-1				
0				
1				
2				
3				

8. Do you see any relationship between the value of the third difference and the value of the coefficient of x^3? (This will be a new relationship, not the same one you saw with the quadratic functions.)

9. At this point, you might feel ready to venture a theory. For a polynomial function of degree n, how many successive differences must we take before arriving at differences that are constant? Test your theory on any polynomial p of degree 4.

x	$p(x)$	First differences	Second differences	Third differences	Fourth differences
−3					
−2					
−1					
0					
1					
2					
3					

10. Test it again with a polynomial q of degree 5.

x	$q(x)$	First differences	Second differences	Third differences	Fourth differences	Fifth differences
−4						
−3						
−2						
−1						
0						
1						
2						
3						
4						

11. You realize, of course, that a bunch of examples does not constitute a mathematical proof. Nevertheless, what you have seen is true: for a polynomial of degree n, the nth differences are constant. What's more, you can predict the value of that n^{th} difference from the polynomial itself. Here's how: compute $n!$. (The symbol $n!$ is pronounced "n factorial," and it means $n \cdot (n-1) \cdot (n-2) \cdots 2 \cdot 1$, so that $4!$ is $4 \cdot 3 \cdot 2 \cdot 1$, or 24.) Multiply $n!$ by the coefficient of x^n in the original polynomial, and you should obtain the value of the nth difference. Check it out for the examples you already did in this investigation.

5.4: GONE TO POT

In a small ceramics studio, the total cost of a day's production depends upon how many pots are produced. When only a small number of pots is manufactured, the per-item cost is high, and total production costs rise rapidly with each additional pot produced. As production increases, efficiency reduces the cost per item, so the *rate of change* in the total cost of production decreases. If the studio attempts to overproduce, however, the extra associated costs cause the rate of change in the total cost to rise again.

Suppose the total cost of operating the studio on a day on which x pots are produced is given by the function

$$C(x) = 0.01x^3 - 0.65x^2 + 14x + 20$$

where C is measured in dollars.

1. Draw a graph of $C(x)$ over the interval $0 \leq x \leq 40$. Scale the axes and label them.

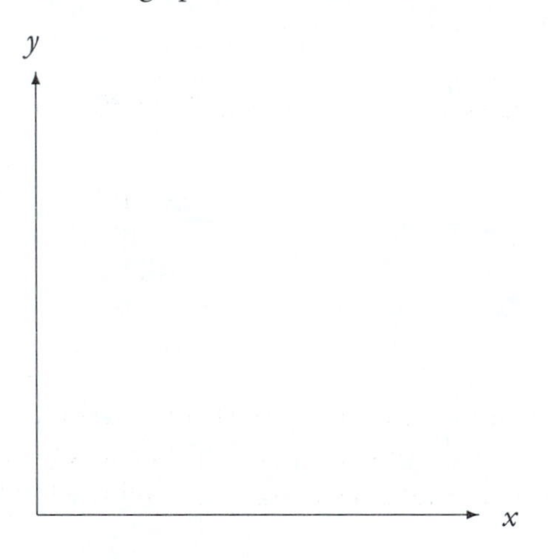

2. What is the fixed cost? (Fixed cost means the expenses incurred even if nothing is produced.)

3. In general, as more pots are made, the total cost increases. Is that always the case for the function $C(x)$? Carefully examine the graph of $C(x)$ and support your answer with specific production levels and their associated costs.

4. Calculate the average rate of change in total cost when production is increased from 1 pot to 3 pots. Give the units for this rate.

5. Calculate the average rate of change in C when x changes from 20 to 22. This result should either confirm your answer to (3) or cause you to rethink your response.

6. Calculate the average rate of change in C when x changes from 36 to 39.

7. Describe what happens to the rate of change in $C(x)$ as x goes from 0 to 40. Do not describe the function itself. Words to use in your description of the rate of change include "positive," "negative," "increasing," and "decreasing." You will need at least three separate statements.

Lab 6: Doormats

I myself have never been able to find out precisely what feminism is:
I only know that people call me a feminist whenever I express sentiments that differentiate me from a doormat.

—Rebecca West, 1913

PREPARATION

Prepare for this project by reading the lab pages and writing formulas for the three average-cost functions F, G, and H. Bring the formulas with you to the lab, and begin by comparing your functions with those of your partners.

$F(x) =$ ______________________________

$G(x) =$ ______________________________

$H(x) =$ ______________________________

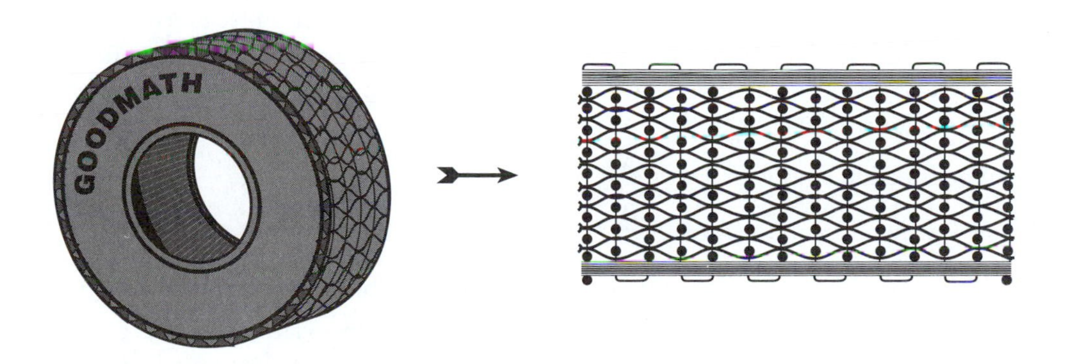

THE DOORMATS LAB

A western Massachusetts couple runs a small cottage industry from their home. They recycle discarded automobile tires by slicing them up into strips and making doormats, somewhere between 500 and 2000 doormats each year. Their somewhat imprecise method of cutting the tires results in a product of irregular thickness, perfect for scraping off mud and snow from shoes. Worn-out tires are a disposal problem nationwide, so they have no difficulty acquiring their raw materials. Additional materials are few and inexpensive—some colored plastic beads for spacing and some thick aluminum wire for holding everything together.

The couple has figured out that, counting the cost of the raw materials and the value of their labor, it costs them \$5 to produce each doormat. In addition, they estimate that their business has fixed annual expenses of \$200. (This is a very "low-tech" operation.) In the language of economics, the capital cost is \$200 and the variable cost is \$5 per unit. They want to determine the average cost of each mat they make this year.

Keeping it Simple

First write a formula for the total cost of producing x doormats this year. Now write a function that expresses the average cost of a doormat. Using a grapher, explore the behavior of this function, which we'll call F. Is the average cost a constant? If so, what constant? If not, why not?

What's the average cost if they produce 200 doormats this year? 500? What happens to the average cost if they produce very few?

As the number of doormats produced increases, what happens to the average cost? (Economists call this effect "spreading the overhead.") Is it reasonable to suppose that, as more and more doormats are made, the average cost would continue to drop? If you study the graph for large values of x, you should see that the curve appears to level off. Is there a lowest possible average cost? Verify that, although the values of the function get smaller and smaller, there's a limit below which the average cost cannot go. From the graph, determine this limit. Does the function ever attain that value? In other words, if we call the limit L, can you find a value of x such that $F(x) = L$? Give a reason for your answer.

Enlarge the x-interval of the viewing window so that it's at least 10,000 units wide. You may include negative values of x if you wish. The graph of $F(x)$, except near the y-axis, should resemble a horizontal line. Give the equation of that horizontal line. Because the graph of $F(x)$, in the long run, resembles the line, we call that line the **horizontal asymptote** for the function F.

The function F as you've written it probably has the form $F(x) = \dfrac{a + bx}{x}$. Rewrite it now (by dividing) in the form $F(x) = \dfrac{a}{x} + b$.

If the magnitude of x is very large, what can you say about the value of the term $\dfrac{a}{x}$? Explain how your answer to that question shows that, for large values of x, the graph of $F(x)$ resembles the line $y = b$. We say that the graph of $F(x)$ has the horizontal line $y = b$ as an asymptote, that the function approaches the value b asymptotically.

View the graphs of $F(x)$ and $y = b$ together so that you can appreciate their similarities for large values of x.

The More the Merrier

The doormat makers have discovered that they become more efficient economically if they make larger numbers of mats. Perhaps they can get a better price on the beads, or perhaps their enthusiasm propels them to work more quickly. You can express this efficiency mathematically by including a quadratic term in the *total cost* function. (Be sure to use *total cost*, not *average cost*.) Suppose that term has been determined to be $0.0003x^2$.

Rewrite the total cost function, to include the quadratic term. (You will need to decide whether to add or subtract this term.) Now, write a new average-cost function; call it G.

Use a grapher to compare the behaviors of $F(x)$ and $G(x)$ over several different intervals, in a variety of viewing windows. Find intervals over which they look nearly the same. Find out approximately where they begin to diverge. At what number of mats does the x^2-term appear to "kick in" and become significant? (This is a matter of opinion; your answer will depend upon how you define *significant*.) Look at the formulas for the two functions and explain why the graphs are very close for certain values of x.

If you use sufficiently large values of x, you should observe that the graph of $G(x)$ begins to look like a straight line rather than a curve. This line, though, should appear slanted rather than horizontal. It is called an **oblique asymptote** or **slant asymptote** for the function G. Here are two methods for discovering an equation for that line.

1. Using a very large interval centered at 0, such as $-10{,}000 < x < 10{,}000$, examine the graph of $G(x)$. Except for some bizarre behavior near the y-axis, it should resemble a straight line. Since you can write an equation for a line if you know two points, use the trace feature to find the coordinates of a point at the left side of the screen and of another one at the right side. Write the equation in $y = mx + b$ form.
2. As you did with the first model $F(x)$, rewrite the formula for $G(x)$ by dividing, so that it now has the form $\frac{a}{x} + b + cx$. Think about what happens to each term as x grows very large. The expression has three terms, but one of them is so close to zero when x is very large that its impact on the function G is negligible. Compare the results with those you obtained with the first method; they should be similar, though not identical, because the first method gives only an *approximation* to the slant asymptote. Do you see now how you can use the formula for the function to determine its asymptote? This method probably took much less time than the first method; use it from now on when you want to determine the long-range behavior of a model.

What does this have to do with making doormats? Go back to some realistic values for x and interpret the behavior of the graph in the context of this situation.

Too Much of a Good Thing

The doormat makers have also observed that, if they spend all of their energies making doormats, their productivity declines, and therefore the cost to them begins to rise again. You can express this mathematically by including a cubic term in the *total cost* function. Decide whether the term ought to be added or subtracted.

Suppose the cubic term has been determined to be $0.0000001x^3$. Rewrite the total cost function to include this term. (Hang onto the quadratic term as well.) Find a third average-cost function, $H(x)$, which includes the cubic term. Then use the grapher to examine all three functions over an interval that makes sense in context. Does the graph of $H(x)$ suggest an optimal number of doormats for the doormat makers to produce?

Now investigate the long-run behavior of $H(x)$. Enlarge the viewing window so that the dramatic behavior near the y-axis is minimized. You'll probably need an x-interval of 40,000 units or so. In this large viewing window, what is the apparent shape of the graph of $H(x)$?

Check this out algebraically: rewrite the formula for $H(x)$ by dividing, as you did in the last two sections. How many terms are there? Which one is of negligible size whenever x is a huge number? Explain why, for enormous values of x, the graph of $H(x)$ looks the way it does.

The function H has an asymptote that is neither horizontal nor oblique; its asymptote is another curve. Give the formula for that curve. (Remember that you can obtain it from the formula for $H(x)$.) Overlay the graph of the curved asymptote, try several different windows, and observe that the graph of $H(x)$, from a distance, resembles the graph of its asymptote.

So, what about doormats? Interpret the asymptotic behavior of the function H in the context of this cottage industry.

THE LAB REPORT

You are serving as mathematics consultants to the doormat makers, who are going to select one of the three average-cost models. Tell them about the three different models. Explain what each term in the function represents and why the term is positive or negative. Compare and contrast the functions for their ability to model a concrete situation. Your role is to explain objectively what each model says, not to tell them which one to use.

Describe the long-term behavior of each of the functions. Include graphs. Show how the algebraic formula for each function can be used to predict its asymptote.

In the light of these long-term behaviors, conclude with three separate recommendations for the doormat makers. What might they want to adopt as production goals if they were to use model F? What advice would you give them if they decided that G were a more accurate model for their business? And finally, how many mats would you recommend they produce if model H appeared to be the best fit for their operation?

Section 6—Projects and Explorations

6.1: TAKING THE PLUNGE—Vertical Asymptotes and Holes

Il est donc possible que les plus grands corps . . . de l'univers, soient . . . invisibles.
—Pierre Simon, Marquis de Laplace

It is therefore possible that the largest bodies . . . of the universe are . . . invisible.
—Authors' translation

In the Doormats lab, you explored what happens to a rational function in the long run—that is, when the magnitude of the independent variable gets very large. In this assignment, you will investigate the behavior of the graph of a rational function in the vicinity of a number that isn't in the domain of the function.

1. Using a grapher, view the graphs of several functions having the form

$$f(x) = \frac{x^2 - k}{x - 2}$$

   (You choose values for k.) Experiment with the viewing window until you are able to see two distinct branches for each of the graphs (or for all but one of the graphs). Your best viewing window will probably be one that's only about six or eight units wide. (Reminder: enclose the numerator and the denominator in parentheses.)

2. Notice that 2 is not in the domain of any of these functions. Arrange the viewing window so that $x = 2$ is at the center and observe the dramatic behavior of the curves as they approach $x = 2$. You should see that they become nearly vertical. Why? What is happening to the denominator of each function as the value of x approaches 2? Explain how the size of the denominator affects the value of the functions.

   The line $x = 2$ is called a **vertical asymptote** for those functions. The closer x gets to 2, the more the graphs actually resemble that vertical line.

3. You might have noticed that what was just stated is not entirely true! Was this one of your functions? If not, graph it now.

$$g(x) = \frac{x^2 - 4}{x - 2}$$

What's going on here? Why does the graph of $g(x)$ look so different from all the others? Examine the formulas, and explain the algebraic difference between $g(x)$ and the other functions you graphed.

4. If you look closely at the screen, you might be able to find a hole in the graph of $g(x)$ at $x = 2$. Even if you don't see any hole, you know that the function isn't defined for $x = 2$, and so, despite what the grapher seems to indicate, the graph actually stops on one side of 2 and starts up again on the other side. Except for the hole, to what simpler function is $g(x)$ equivalent?

5. Sketch a graph of $g(x)$. How will you indicate that the number 2 is not in its domain? Label the coordinates of the hole.

6. Find the formula for another rational function that has a hole. Sketch its graph, labeling the coordinates of the hole.

7. Sketch the graph of $f(x) = \dfrac{x+2}{x^2+x-2}$, showing clearly any vertical asymptotes or holes. (*Hint:* this function has one of each; factor the denominator and examine the factors carefully.)

6.2: FROM HERE TO ETERNITY—Long-Term Behavior of Rational Functions

In the Doormats lab, you studied the graphs of three rational functions. You saw that, for very large values of the independent variable, the graph of each rational function resembled the graph of a simpler function—a line or a parabola. In this assignment, you will investigate several other rational functions and learn a method for predicting their nonvertical asymptotes—that is, the simpler functions they eventually resemble.

A **rational function** is defined as the quotient of two polynomial functions, where the polynomial functions can be as simple or as complicated as we wish. In the Doormats lab, the denominator for each of the rational functions was the very uncomplicated polynomial x, and it was relatively easy to figure out what those graphs were going to look like in the long run.

1. Define a function that has the form $\dfrac{ax + b}{cx + d}$—that is, a linear expression divided by another linear expression. (You choose values for a, b, c, and d.) What sort of asymptotic behavior might be expected? Investigate the graph of your function in a very wide viewing window. Does it level off? Does it resemble a curve? If you've typed the function correctly, you should see the graph leveling off at both sides of the screen and beginning to resemble a horizontal line. Use the graph to write an (approximate) equation for the nonvertical asymptote, which in this case will be a horizontal line.

Your function	Its asymptote

2. Now get some help from algebra: using long division, divide the denominator into the numerator. You should get a *number* plus a remainder. In other words, you've shown that

$$\frac{ax + b}{cx + d} = n + \frac{m}{cx + d}$$

where m and n are other constants.

When x is very large, what can you say about the quantity $\dfrac{m}{cx + d}$? If you weren't sure about the asymptote before, you should be able to read its equation from the calculations you've just done.

3. Choose two other functions that have the same form (linear over linear). Predict their nonvertical asymptotes. Use a grapher to check. Are all of those asymptotes horizontal?

Your functions	Their nonvertical asymptotes

4. Try a function that's a quotient of two quadratic expressions, such as

$$\frac{2x^2 + 3x - 1}{3x^2 + 2}$$

Could you have predicted its asymptote simply from inspecting the formula?

5. Experiment with other forms of rational functions. Try one that has a linear expression on the top and a quadratic on the bottom. Do you have any expectations about the nonvertical asymptote?

Try dividing the bottom into the top. You're stuck, aren't you? Why? This has implications for the nonvertical asymptote. Graph the function, and describe how it looks for large values of x. What does this tell you about the asymptote?

Your function	Its nonvertical asymptote

6. Look for a pattern by trying a couple more functions in which the bottom has a higher degree than the top. They all have the same horizontal asymptote. What is it?

Your functions	Their nonvertical asymptotes

7. Now switch tops and bottoms, so that the top has a higher degree than the bottom. Try to figure out what's happening, both algebraically and graphically. For the graph, you need only look—in a very wide viewing window. For the algebra, you need to perform long division and think about what portion of the result is negligible when x becomes very large. Be sure that you're doing some examples that are other than a quadratic over a linear; you should include some expressions of higher degrees and you should include some functions in which there's at least a two-degree difference between the top and the bottom. (Aren't you glad you don't have to draw all these graphs by hand?)

Your functions	Their nonvertical asymptotes

8. After enough exploration, you should be ready to draw some conclusions and write a summary of the long-term behavior of rational functions. There are really only three separate cases: (a) the degree of the numerator and the degree of the denominator are the same; (b) the degree of the numerator is less than the degree of the denominator; (c) the degree of the numerator is greater than the degree of the denominator. Tell what you learned about each case.

   (a)

   (b)

   (c)

9. *Globally, the graph of a rational function resembles that of a polynomial.* (Recall that constant functions and linear functions are special cases of polynomials.) Tell how your work in this assignment would support that statement.

6.3: LET'S BAN WASTE!

In recent years, people concerned about the environment have campaigned to encourage moderation and conservation in our use of resources. The U.S. Public Interest Research Group (PIRG) presents "Wastemaker Awards" to manufacturers who, it claims, use unnecessary amounts of packaging for some of their products. Health and beauty aids appear to be among the worst offenders, winning all seven of the 1991 awards.

One of these dubious honors went to Bristol-Myers Products for its Ban Roll-On deodorant. One-and-a-half fluid ounces (44 cubic centimeters) of liquid come packaged in a plastic container, mainly cylindrical in shape but pinched in the middle. The cylinder is 5.1 cm high and has a diameter of 3.8 cm. It is topped by a plastic cap 4.1 cm high, having the same diameter. All of this rests in an outer container of cardboard, a rectangular box whose dimensions are 13.3 cm by 7.0 cm by 4.1 cm. There is, in addition, some inner cushioning, also made of cardboard.

1. Calculate the approximate surface area of the packaging materials used for a 1.5 ounce container of Ban Roll-On. How much is cardboard? (You may ignore the inner cushioning as well as any overlapping of flaps that would be found in a box.) How much is plastic? (Again, you may ignore the region where the cap overlaps the main container.) The plastic would be tricky, because the shape isn't a perfect cylinder, so you'll have to pretend that it is one. Do you think your your answer is an overestimate or an underestimate of the actual amount of plastic used? Explain.

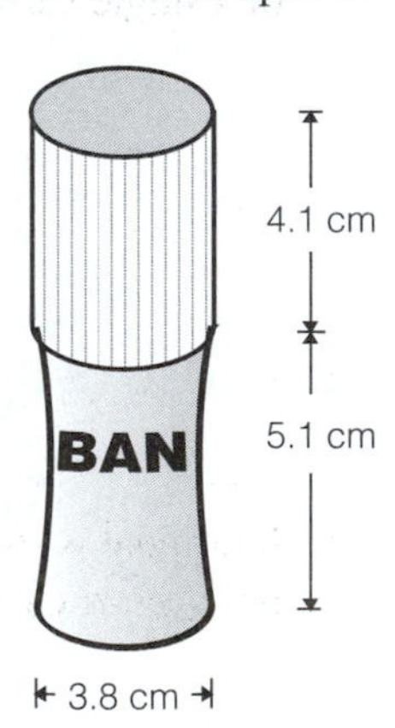

2. Now you will figure out just how efficient Bristol-Myers might have been with their plastic container. Suppose you wanted to package 44 cubic centimeters of liquid in a cylindrical container. (We're going to ignore details such as providing a way to open the package.) Write an equation expressing the surface area S as a function of r and h. Remember to include the top and the bottom.

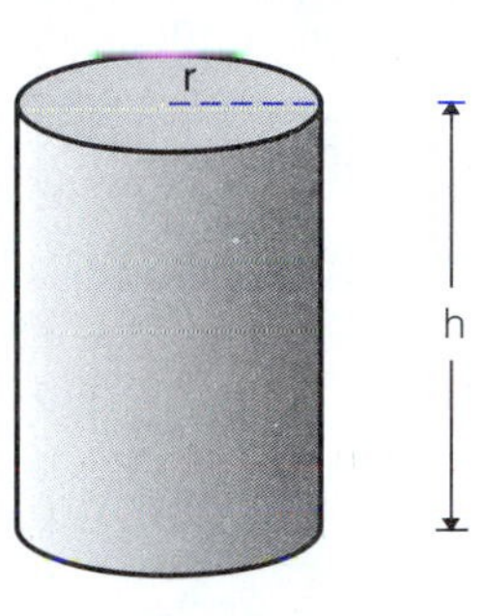

3. Express the volume V as a function of r and h.

4. Since you have determined that V is to be 44 cubic centimeters, use 44 for V. Now you can solve for one of the variables in terms of the other. Substitute that expression into your surface-area function. The surface-area function should now be a function of a single variable. What is the independent variable?

5. Your goal is to minimize the surface area, and thus the amount of plastic. Can you tell from the formula alone whether the function has a minimum? Why or why not?

6. Use your grapher to graph the function. Choose a sensible interval for the independent variable and look for a minimum in that interval. (You should see a curve that dips down and then rises. If your graph doesn't have the right shape, check to see whether you've enclosed the denominator in parentheses.) Find the coordinates of the turning point, accurate to three decimal places, and use them to determine the best value of the independent variable. Make a sketch of the relevant portion of the graph, labeling the coordinates of the minimum. Put scales on the axes and label them.

7. Give the dimensions (r and h) of the most efficient cylinder and tell how much plastic it uses (surface area).

8. What is the ratio of h to r?

9. Note your answer to (8). That ratio is not any old random result, but an example of the elegant patterns to be found in mathematics. In fact, for a cylindrical container of *any* size, we can minimize its surface area by making the height twice as large as the radius. Or, to put it another way, by making the height equal to the diameter. If you were to hold such a container upright in front of a light, what would be the shape of the shadow it would cast on an opposite wall? Does this remind you of Jack's field from the Galileo lab?

10. Conserving materials doesn't appear to be one of the major concerns of this manufacturer. In designing the package, what might be some other considerations? Do you think your optimal cylinder would serve its purpose well? Do you think the cardboard is necessary or important? Give reasons.

6.4: DRUG TESTING

How reliable is a drug test?

The concern over increasing drug usage among teens has prompted some schools to consider mandatory drug testing of its students. In addition, worried parents now can administer relatively inexpensive tests to their own children. But caution is in order. Drug tests, while generally reliable, are not perfect. Sometimes a person who does use drugs gets a negative result (a **false negative**). Sometimes a person who does not use drugs tests positive (a **false positive**). Companies who produce tests for drugs usually provide information on two characteristics of their test: p, the probability that their test correctly identifies a drug user; and r, the probability that it correctly reports the absence of drugs. The numbers p and r are decimal values between 0 and 1, with 1 representing a probability of 100 percent.

A large high school in your community is considering a mandatory drug testing program. The local school board has asked your mathematics instructor to quantify the benefits and the dangers of instituting such a program. The board is particularly concerned that the percentage of positive tests from nonusers might be too high. Applying principles of conditional probability, your instructor derived this relationship between E, the percentage of positive test from users, and U, the percentage of drug users in the group being tested:

$$E = \frac{100\,p\,U}{(p + r - 1)U + 100(1 - r)}$$

You, the members of her precalculus class, are to investigate the implications of this function and prepare a summary for the school board.

1. The school is considering a test that correctly identifies 95% of the users ($p = 0.95$) and correctly reports the absence of drugs in 90% of the nonusers ($r = 0.90$). Write the function describing the relationship between E and U for this drug test.

2. Graph that function, using a window that shows the key features of the *abstract* function. Sketch the graph, identifying the horizontal and the vertical asymptotes.

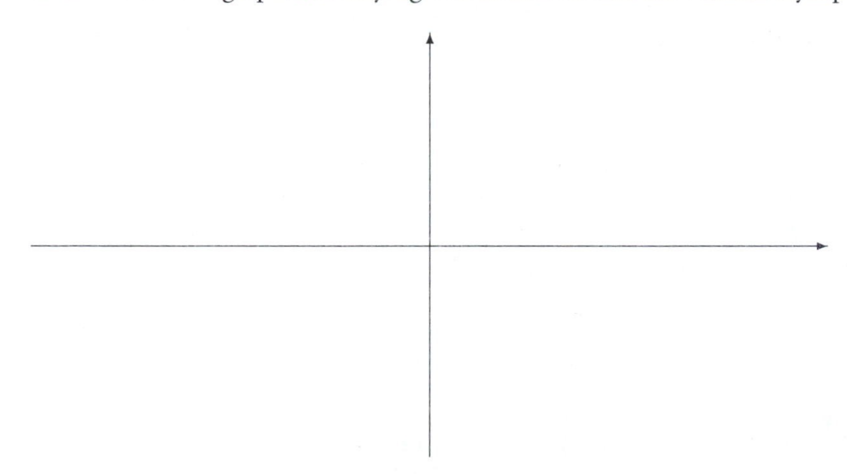

3. Highlight the portion of the graph that represents the drug-test model, recalling that E and U are *percentages.* Does either of the asymptotes you identified have meaning in the context of drug testing? Explain.

4. Suppose that 25% of the students are drug users ($U = 25$). What percentage of the positive tests will be from users? Show an algebraic solution, and explain how your graph confirms the answer.

5. Suppose instead that only 2% of the students use drugs ($U = 2$). What percentage of the positive tests will be from users? What does this imply about the rest of the positive tests?

6. The answers to (4) and (5) show that there would be a high percentage of false positives—positive test results for students who do not use drugs. The committee, therefore, is considering a more expensive but more reliable test, whose characteristics are $p = 0.99$ and $r = 0.95$. Analyze the more expensive test if 25% of the students are users and again if 2% are users.

7. Compare and contrast the results, and summarize your findings in a presentation you could make to the school board.

8. The function your instructor derived applies equally well to medical tests, such as the one for HIV, the virus that causes AIDS. Tracking people infected with HIV might be helpful in controlling the spread of the disease. One particular test correctly detects the presence of HIV antibodies 99.7% of the time and correctly reports the absence of HIV antibodies 98.5% of the time. Do you think the government should institute a program of mandatory testing for HIV? Explain why or why not, and support your answer mathematically.

   In order to use the function, you will have to estimate U, the percentage of the population that is HIV-positive. Be sure to state your assumptions.

You might have found some of the results in this project to be startling or disturbing. The function E shows that, even with a highly reliable test, the percentage of true positives will be relatively low and the percentage of false positives alarmingly high whenever the condition you're testing for is one that affects only a small percentage of the population. For this reason, among many, the issue of mandatory drug testing is controversial.

Lab 7A: AIDS

Bridge over troubled water

PREPARATION

In the 1980s, the United States had the dubious distinction of being the first country to experience the devastation of the AIDS epidemic. In fact, 10,805 of the 13,170 cases of AIDS reported to the World Health Organization through 1984 occurred in the U.S.[1] Distress over the tragedies of individual cases was compounded by the realization on the part of public health officials that a mathematical model for the spread of the epidemic was an exponential function. In an essay in the *New York Times* (April 19, 1987), the noted biologist Stephen Jay Gould wrote, "The exponential spread of AIDS underscores the tragedy of our delay in fighting one of nature's plagues."

To appreciate the urgency of the situation described by Gould and others, we need to understand the nature of exponential growth. The simplest example is a trick problem that you may have seen before:

> If you place a penny on the first square of a checkerboard, two pennies on the second square, four on the third square, and eight on the fourth, and if you continue doubling the number of pennies on each subsequent square, how many pennies are on just the last (64th) square?

Write the answer using an exponent, and then give the dollar value of the pennies. (*Note:* If you could stack the pennies, you'd have a stack about as high as the universe is wide!)

Fill in the following table, in which the second column gives the number of pennies on the square whose number is in the first column.

Number of square	Number of pennies
1	
2	
3	
4	
5	
6	

Graph the points determined by the table. This is an example of **discrete** data; the only values that are meaningful for both the x- and the y-coordinates are positive integers. Now, to more easily perceive the pattern of the data and to help determine the mathematical function that can model the data, connect the points with a curve as smoothly as possible.

Letting y be the number of pennies on the xth square, write an equation giving y as a function of x. The function you determined is an **exponential** function. In this function, where does the independent variable appear?

Make four more graphs, one for each data set given in the AIDS lab. Bring all the graphs with you to the lab.

[1]J. M. Mann and D. J. M. Tarantola (Editors), *AIDS in the World II*, Oxford University Press, 1996.

THE AIDS LAB

The home page on the World Wide Web for the Centers for Disease Control in Atlanta shows graphs of death rates in various sectors of the U.S. population. In 1992, HIV infection became the leading cause of death among men aged 25–44 years; in 1994, it became the third leading cause of death among women in that same age group. (HIV is generally considered to be the virus that causes AIDS.) Many cases of AIDS in the heterosexual population have been associated with intravenous drug use. The following table, however, gives the much smaller numbers of cases of AIDS among women in the United States ages 13 and older who were neither intravenous drug users nor sexual partners of users of intravenous drugs.[2]

AIDS in U.S. Women (cases not associated with intravenous drugs)

Year	Cases reported
1982	2
1983	7
1984	13
1985	31
1986	84
1987	151
1988	226
1989	330
1990	503
1991	726
1992	972

Plot the points corresponding to this table. Let the integers 0, 1, 2, ..., 10 on the horizontal axis correspond to the years 1982, ..., 1992 and the numbers on the vertical axis represent the number of AIDS cases reported in this population of women. Draw a smooth curve that approximates the data (don't attempt to hit all the points) and use the curve to estimate the number of cases reported in 1993.

In the exponential function that we used to model the coins on a checkerboard, the *base* was 2, because the number of coins *doubled* with every square. Compare with your partners the formulas each of you wrote for the checkerboard problem. When you are in agreement, use your grapher to draw the function you determined. Experiment with viewing windows until the graph on the screen matches the one you drew.

Now let's return to the AIDS statistics. That graph should have the same general shape as the one you drew for the checkerboard problem. If your grapher has a curve-fitting command, ask it to fit an exponential function to the data. You should get a function like this one:

$$f(x) = 4.244(1.825)^x$$

That function, $f(x) = 4.244(1.825)^x$, is given in standard exponential form,

$$y = C \cdot b^x$$

[2]W. A. Rushing, *The AIDS Epidemic*, Westview Press, 1995, p. 109.

where b, the constant base, is 1.825 and C, the constant coefficient, is 4.244. (The base for the checkerboard problem is 2. That function, however, unless you did some algebraic fiddling, is not in standard form because its exponent is not a simple unadorned x.)

Graph the exponential model together with the data points. Do you think the curve represents the data well? The model is a sort of *average* of the data. Although some points are above the curve and others are below, as we should expect, observe that the too-high points are clustered in the middle and the much-too-low points appear on the right. These details suggests that the overall trend of the data is about to change.

We might need something more sophisticated than an exponential function if we are to model this population for years beyond 1992. We'll return to this idea later in the lab and again in Project 7.7, *Limits to Growth*. For now, though, let's stick with the model we have.

In the checkerboard problem, each time the input variable increased by one square, the output (number of pennies) was multiplied by 2. For the exponential model $f(x)$, each time the input variable increases by one year, the output (number of cases) is also multiplied by a specific amount. What is that amount?

The coefficient C for the women's AIDS model is larger than the coefficient C for the checkerboard model. This causes the AIDS model to give larger output values at first (for small values of x) than the checkerboard function does. However, no matter how great the difference in C values, the function with the larger base will eventually overtake the function with the smaller base. Which model has the larger base?

Use your grapher to estimate the first integer value of x for which the checkerboard model is greater than the AIDS model. (*Hint:* it's greater than 20.) Both of these curves become incredibly steep as x grows. If you have trouble reading values from the graphs, you could instead evaluate each function for several values of x starting at 20, and notice how the checkerboard model catches up to the AIDS model.

The first point that you plotted for the AIDS data was $(0, 2)$, although that point does not lie on the graph of the model $f(x) = 4.244(1.825)^x$. What is the value of the model when $x = 0$? When $x = 1$? The points that you plotted all have integer coordinates, but the function that models the data assigns rational and even irrational values of y to integer values of x. Remember that a model only *approximates* the data. Even as far back as the Fahrenheit lab, you created a model of the temperature sign data that didn't give the exact data points. In this lab, we'll look at other reasons that a model might fit the data imperfectly, and then over only a limited time interval.

Now let's look at another set of data. In an article by R. Steinbrook in the *Los Angeles Times* from January 5, 1990, "Slower Spread of AIDS in Gays Seen Nationally," the following information on the number of AIDS cases reported in Los Angeles was given:

AIDS in Los Angeles

Date of diagnosis	Cases reported
Jan.–June '83	116
July–Dec. '83	154
Jan.–June '84	197
July–Dec. '84	269
Jan.–June '85	415
July–Dec. '85	503
Jan.–June '86	668
July–Dec. '86	773
Jan.–June '87	952
July–Dec. '87	933
Jan.–June '88	955
July–Dec. '88	943
Jan.–June '89	967

To get a feeling for these data, plot the points corresponding to the table, numbering the horizontal axis 0, 1, 2, . . . to count the time intervals, with 0 representing the period January–June 1983, and number the vertical axis to count the AIDS cases reported in Los Angeles during that period. You will see that, in this case, not all of the points appear to lie along an exponential curve. To create a model for this situation, we will need to splice two functions together. Here's a start. Using your grapher to help you, draw the graph of the exponential function

$$j(x) = 121(1.313)^x$$

on the same axes where you plotted the points from the table.

For what time period is j *not* a good model? For those x-values, a much simpler function will serve to model the data. Give such a function and the interval on the x-axis where it is appropriate. Now use both formulas to write a *single* piecewise function that could represent the entire period 1983–1989. Be sure to indicate which formula goes with which time period.

The article in the *Los Angeles Times* stated that "public health officials cite several possible causes for the slowdown, including the adoption of safer sexual practices by many gay men to prevent infection." In this situation, changing behavior altered the basic conditions contributing to the epidemic and, fortunately, made the exponential model obsolete for these data. (Go back and check the date of the article by Stephen Jay Gould that introduced this lab.)

There is another reason why an exponential function can model a real-world situation over only a brief period. Since exponential functions grow so rapidly, they exhaust the population very quickly.

A common pattern for an epidemic is exemplified by the following table, which represents data from the entire U.S. population. We rounded values from *The AIDS Epidemic*[3] to the nearest thousand cases.

AIDS in the United States

Year	Cases diagnosed (in thousands)
1981	1
1982	1
1983	3
1984	6
1985	12
1986	19
1987	28
1988	34
1989	39
1990	41
1991	42

Plot the points corresponding to this table on a graph where the integers 0, 1, 2, . . . , 10 on the horizontal axis correspond to the years 1981, . . . 1991, and the numbers on the vertical axis represent AIDS cases (in thousands) diagnosed in the United States that year. Draw a smooth curve that approximates your data. Color the part of the graph that looks like an exponential curve. After approximately what year does the exponential function no longer provide a good model? (By the way, in what year did Stephen Jay Gould write the essay mentioned in the lab preparation?)

It is more difficult to write a function for this graph, and we won't ask you to try! The points follow, approximately, a pattern that mathematicians call **logistic growth**; you'll see a way to model logistic growth in Project 7.7.

Although the epidemic has slowed in the United States, it continues to rage unchecked in other parts of the world, particularly in sub-Saharan Africa and southeast Asia. An article in the July 5, 1997, issue of *The Economist* states that the number of South Africans infected with HIV is 2.4 million, three times as many as in the United States. Ironically, the article suggests that the ending of apartheid has hastened the spread of the disease by opening the South African economy to greater trade and migration flows from the north.

The following table gives the number of AIDS cases in South Africa as reported in *AIDS in the World II.* (These are cases of AIDS, not simply of HIV infection, and include only those reported to the World Health Organization, WHO. Many cases of the disease go unreported.)

[3]Rushing, op. cit., p. 111.

AIDS in South Africa

Year	Cases reported to WHO
1985	9
1986	34
1987	48
1988	94
1989	176
1990	304
1991	393
1992	658
1993	1267
1994	2774

Graph this set of points as you did the others in this lab. Label the x-axis 0, 1, 2, . . . , 9 to correspond to the years 1985–1994, and the vertical axis to count the number of AIDS cases reported. Which of the other three graphs in this lab does it most nearly resemble? If your grapher has a curve-fitting command, ask it to approximate the data with an exponential function. You should get a model similar to

$$h(x) = 14.437(1.777)^x$$

The base of this standard-form model tells the amount by which the output from a given year is multiplied to obtain the output for the following year. What is that multiplier?

We can think of that multiplier as a percentage. The base, 1.777, can be written as 177.7%, or 100% + 77.7%, indicating that AIDS in South Africa is growing, on average, at 77.7% per year! Look again at the first model, $f(x) = 4.244(1.825)^x$, and figure the annual percentage growth of AIDS in that population of U. S. women.

THE LAB REPORT

Write a report explaining to a student who's not taking your precalculus course why Stephen Jay Gould could state (in the same article quoted in the lab preparation), "The AIDS pandemic . . . may rank with nuclear weaponry as the greatest danger of our era." In your report, describe the nature of exponential growth and tell how the base of the exponential function governs the growth of the function values. Place Gould's essay in historical context by noting the year in which it was written and relating it to the information conveyed by each of the four graphs. Then describe the limitations inherent in an exponential model, referring to the data for the United States and for Los Angeles. Give the complete function that you wrote to model the Los Angeles data. Finally, discuss the alarming nature of the women's AIDS data and the statistics from South Africa, noting their similarities and what they indicate about the stage of the epidemic in these two population groups. Include graphs of the four sets of AIDS data and of the three mathematical models.

Lab 7B: Radioactive Waste

Waste is a terrible thing to mind

PREPARATION

From 1943 to the late 1950s, the United States operated a secret scientific enterprise for developing materials for atomic weapons that covered more than 100 sites in 32 states and the Marshall Islands in the South Pacific. Since 1988, the Department of Energy has struggled to account for the astonishing array and volume of radioactive wastes that each site produced.

From South Carolina to Alaska, the program to clean up wastes left by the atomic weapons industry has become one of the Government's most technically difficult and expensive problems. The program is expected to cost $200 billion . . . .

Nearly a million cubic yards of dirt contaminated with thorium, uranium and radium, enough to fill much of Shea Stadium, are spread across abandoned processing plants in two sites near Buffalo and seven cities in New Jersey. More than two million cubic yards of spoils contaminated with radioactive elements and toxic chemicals are piled in five dumping grounds in the St. Louis area. In the West, enormous piles of radioactive sand produced by uranium processing mills are waiting to be sealed beneath tons of dirt and desert stones.[1]

—"In the Trail of the Nuclear Arms Industry,"
by K. Schneider, *The New York Times,* August 26, 1990

Radioactive substances decay over time by emitting alpha and beta particles and gamma rays. The Nuclear Regulatory Commission (NRC) uses special computer software that holds information on hundreds of radioactive substances and will model various conditions, such as ground water and soil types, at storage sites in an effort to manage the containment of radioactive substances responsibly. Have you ever questioned how long radioactive materials would need to be cared for before they are safe?

The **half-life** of a radioactive substance is the time it takes for half of the substance to decay (into other substances, some of which are also radioactive). Let's assume that someone finds 10 grams of uranium-232 at one of the sites. U-232 has a half-life of 72 years. Thus, after 72 years only 5 grams of U-232 would be left (the missing 5 grams having decayed into other substances or lost as released energy). After two half-lives (144 years) 2.5 grams of U-232 would remain. How much U-232 would remain after three half-lives (216 years)? After four half-lives (288 years)?

Sketch a graph of the relationship between the amount of U-232 and time, using the horizontal axis for time, measured in half-lives. You have just drawn an exponential decay curve for U-232. (In preparation for the AIDS lab, you drew a graph depicting the results of repeated doubling. Now you are sketching the results of repeated halving.) Using your hand-drawn graph, estimate how much of the U-232 remains after $1\frac{1}{2}$ half-lives. How many years are there in $1\frac{1}{2}$ half-lives?

The amount of U-232 present after t half-lives, $A(t)$, has the form

$$A(t) = C \cdot b^t$$

where C and b are constants that need to be determined before you can examine the decay curve with your grapher.

Use the fact that $A(0) = 10$ to determine the value of C. Replace C with this value. Now use the information that $A(1) = 5$ to find the value of b.

Write down the completely determined decay function $A(t)$ and bring it to the lab, along with your sketch of the graph.

THE WASTE LAB

Decay Curves Using Half-Lives

Graph the decay curve for U-232. Adjust the viewing window so that the graph will have the same scale as the graph you drew by hand. Using your grapher, estimate to at least two-decimal-place accuracy the amount of U-232 remaining after $1\frac{1}{2}$ half-lives. (How do you know that you have achieved the desired accuracy?) Compare this estimate with the one you obtained from the hand-drawn graph.

In how many years will only 10% of the original U-232 remain? (Be careful here; your function measures time in half-lives of U-232. You must convert the answer into years.)

When U-232 decays, it converts into other substances, which in turn decay into still other substances. A partial listing of the decay chain is given below.

Nuclide	Half-Life
U-232	72 years
↓	
Th-228	1.91 years
↓	
Ra-224	3.62 days
↓	
Rn-220	55.6 seconds
↓	
Po-216	0.146 seconds
↓	
Pb-212	10.6 hours

Eventually the chain stops with a stable, nonradioactive substance.

What if you started with 10 grams of thorium-228 (Th-228)? Determine the function relating the amount of Th-228 to time, measured in half-lives of thorium-228. Sketch the graph. How does this function compare to the decay curve you drew for 10 grams of U-232? How many years would you have to wait until only 10% of the original sample of Th-228 was left?

Repeat the work of the last paragraph, substituting radium-224 (Ra-224) for Th-228. Do you have to work all the calculations from scratch, or can you rely on patterns you observed for uranium and thorium?

You were asked to find the number of years it would take before only 10% of the original radioactive substance remains. To answer this question, you needed to convert half-lives into years, because t, the independent variable of the function $A(t)$, measured time in half-lives.

Decay Curves Using Years

At times, it is more useful to work with a function that relates the amount of original material remaining to time measured in years, instead of in half-lives. (RADDECAY, a software package used at the NRC for training in methodology for radioactive waste management, is capable of producing decay curves with an independent variable of time measured in years, days, or seconds for hundreds of radioactive substances. The user needs only to specify the substance, time unit, and time duration.)

Let's consider the substance U-232 first. Go back to the graph you sketched by hand. Underneath the half-life numbers on the horizontal axis, write the equivalent measurements in years. You now have a graph of the relationship between the amount of U-232 and the number of years elapsed.

In order to graph this function with the grapher, you need an algebraic expression that describes it. The current formula expresses A as a function of t, where t is measured in half-lives. What you want is a formula that expresses A as a function of x, where x is measured in years.

First, write the algebraic relationship between t and x, solving for t so that you have t as a function of x. Then go to the $A(t)$ formula and substitute the $t(x)$ expression in place of t. You should now have a *composed* function $A \circ t(x)$, whose input x represents time measured in years and whose output is the amount remaining after x years.

Graph the function $A \circ t(x)$. Verify that your function is correct: $A \circ t(72)$ should equal 5 grams. Does it? Use your grapher to determine how much uranium-232 remains after 5 years; after 200 years.

In a similar fashion, construct the function relating the amount of Th-228 to time measured in years (assuming that you start with 10 grams). Graph the function and determine how much Th-228 remains after five years.

Another isotope of uranium, U-235, has a half-life of 703,800,000 years. Suppose that you start with 10 grams of uranium-235. Sketch a graph of the decay curve with time measured in years. (You shouldn't need to use your grapher or even have to write out the decay function explicitly in order to sketch this graph.) Using information from the graph, comment on the implications for safe storage of this substance.

Rates of Decay

You've been studying the relationship between the amount of a specified radioactive substance and the elapsed time. Now let's consider the *rate* at which the substance decays, rather than the quantity remaining. How fast (in grams per year, grams per second, grams per half-life, or whatever units are appropriate) is the substance decaying? Some important questions to consider are these:

- As time passes, does the rate of decay stay the same, increase, or decrease?
- Can you obtain this information by looking at the decay curves?

Let's see how you might find the answers to these questions. Recall that, when you looked at how fast a *linear* function was changing, you saw that its rate of change is constant and is measured by the slope of its graph. Now you will apply this idea to the decay curves.

Start with 10 grams of any radioactive substance. Graph the decay curve with time measured in half-lives. Estimate the rate at which the substance is decaying after one, two, and three half-lives using the following approach. To approximate the rate of decay at, say, $t = 1$, zoom in on the graph at the point $(1, 5)$ until what you see in your viewing screen looks straight. (You will probably need to zoom four or more times depending on the zoom factor of your grapher.) Find the coodinates of two points on this line (write down all the digits your grapher provides) and calculate the slope. Repeat this procedure for $t = 2$ and $t = 3$. Pay attention to the sign of the slope, and state the units in which it is measured.

Does the rate at which the substance decays appear to remain constant, to increase, or to decrease with the passage of time? Do you see a numerical pattern in the sequence of rates that you calculated? Explain.

THE LAB REPORT

How can we use exponential functions to represent the decay of radioactive materials? Explain both functions you developed in this lab and illustrate them with one or more graphs. (A single well-drawn graph with multiple horizontal scales will do the whole job if your labeling is sufficiently clear.) What changes when the independent variable is measured in years rather than in half-lives? How does measuring time in half-lives simplify the mathematics involved in radioactive decay? Compare the length of time it takes for all but 10% of U-232, Th-228, and Ra-224 to decay. Discuss how the rate at which the substance decays depends upon time. Comment on potential problems in storing radioactive substances, backing up this comment with the mathematical evidence you acquired in doing this lab.

Section 7—Projects and Explorations

7.1: POWER TRIP—Exploring Exponential Functions

How can we recognize an exponential pattern? What characteristics do all exponential functions share? In this investigation, you will examine the general exponential function $C \cdot b^x$ to see the effect of changing the constant multiplier C and the base b.

Changing the Base

1. First, we'll find out what b does to the graph. For simplicity, let's start with $C = 1$ and experiment only with the values of b. Try several *positive* numbers for b and make comparisons. Include fractional values—that is, numbers between 0 and 1—as well as bases greater than 1. You should see two distinctly different groups of graphs. Summarize the ways in which the size of the base affects the graph.

2. The graphs you viewed share the following characteristics. Identify them.
   (a) The y-intercept:
   (b) The horizontal asymptote:

3. Let the base be 1, and describe the graph. Compare it with the other graphs you saw. Look at your answers in (2), and suggest a reason why mathematicians do not consider 1^x an *exponential* function. (It's a legitimate function; we just don't call it "exponential.")

4. Suppose the base is 0. What sort of graph do you see? It's possible that you won't see anything at all. If that happens, try viewing the graph with the axes turned off. The graph, in fact, coincides with the right side of the x-axis. Why doesn't the function 0^x exist for $x \leq 0$? Any ideas?

5. Try using a negative number as the base. You'll need parentheses around it, for example, $(-3)^x$. (Without parentheses, the function is simply the opposite of 3^x.) What happens? You probably will not have much success! Substitute several values, including some fractions, for x so that you can get an idea why the grapher is unwilling to draw you a picture.

Changing the Coefficient

6. Now hold the base constant at 2 or 3 and let C vary. Use both positive and negative values for C. Describe its effect on the graph, interpreting this in terms of transformations (stretching, reflection) that you have already learned.

7. What is the y-intercept of any exponential function $C \cdot b^x$?

USING e AS A BASE

Every exponential function $C \cdot b^x$ can be expressed in base-e form: $C \cdot e^{kx}$. There are reasons why we might want to do this. Sometimes it's handy if all the exponential functions we happen to be using have a single base. Also, the number e has some very important mathematical properties, as we will begin to see in 7.2, making functions with this base the easiest to work with in calculus. For now, our goal will be modest: to determine the effect of k on the graph of a function $C \cdot e^{kx}$.

8. Choose a fixed positive value for C (1 is fine), and try a sequence of values for k. Use both positive and negative k's. All functions $C \cdot e^{kx}$, $C > 0$, can be grouped into two classes depending upon k. Explain.

9. Interpret the effect of k as a transformation of the graph of $C \cdot e^{kx}$. What type of transformation does k produce?

10. The constants C and k in a function $C \cdot e^{kx}$ both have the effect of stretching or compressing the graph of e^x, but one effect happens in the vertical direction and one in the horizontal direction. Explain. What feature of the graph would tell you that you were viewing $3e^x$, say, rather than e^{3x}?

7.2: THE INCREDIBLE *e*—Exponential Growth Rates

The number e is used so often in mathematics that it is called the *natural* base for the exponential and logarithm functions. The usefulness of e is most apparent when we apply the techniques of calculus to the study of exponential functions. But you will be able to learn one property of e directly from the graph of the function $y = e^x$.

You have seen that when you zoom in on the graph of a polynomial or rational function, it looks more and more like a straight line. The same is true for exponential functions. (Check it out!)

1. Sketch graphs of the functions 2^x, e^x, and 3^x. Draw them carefully so you can distinguish one from the other, particularly on the left side of the y-axis. (Note that 2^x is *above* the others for $x < 0$.) All three have the same y-intercept. (What is it?) Which of the graphs is the steepest there? Which is the shallowest?

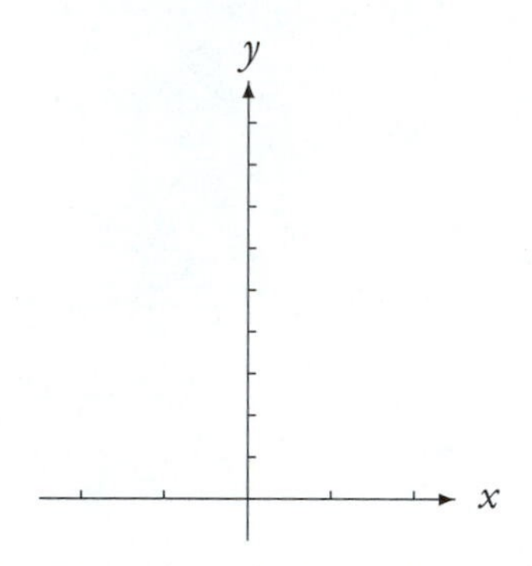

2. We wish to find the slope of each of these graphs at this common point. Zoom in on the y-intercept until the three graphs appear to be straight lines. (Don't "overzoom." A viewing window that's approximately 0.01 or 0.02 unit wide will serve our purpose. Too tight a viewing window can lead to a phenomenon known as "roundoff error.") Locate two points on the graph of 2^x, one on each side of the y-intercept, and use their coordinates to approximate the slope of the curve at the y-intercept. (For accurate results, you should use all the decimal places available in writing the coordinates of the two points and in calculating the slope.) Do the same for the slopes of the graphs of e^x and 3^x at their y-intercepts. When you finish, round off the slope values to two decimal places.

3. What is your best guess for the slope of the graph of $g(x) = e^x$ at its y-intercept? What is the value of g at that same point? In other words, what is $g(0)$? Do you notice something about e^x that is not true of 2^x or 3^x?

4. Choose a different point, not near the y-intercept, on the graph of $g(x) = e^x$. Write down the value of the function g, correct to two decimal places, at that point. Then, as before, estimate the value of the slope of the graph of the function g at that point. (Zoom in; pick two points, one on either side of your chosen point; use all the accuracy the grapher provides; round your final answer to two decimal places.) Compare the slope number to the number you wrote for the value of the function g at that point.

5. Well! Was that last result a fluke? Pick a third point, somewhere else on the graph of e^x, and repeat the slope-finding procedure.

6. Lest you think that we can expect function values and slope values to match each other for other functions as well, try any other function (a parabola, maybe), and show that the value of the function at a particular point is not ordinarily equal to the slope of the graph at that point.

What you have just witnessed is the property of e that makes it unique among numbers:

> For the function e^x,
> the rate of growth at any point
> is equal to the function value at that point.

This is why, whenever we study a process of continuous exponential growth, the number e is involved.

7. You might wonder whether the slopes of the other two functions at their y-intercept have any significance. (What do you expect? This is mathematics—everything has significance!) Using your calculator, write the natural logarithms for 2 and 3 (ln 2 and ln 3). Compare those two numbers with the slopes you already found for 2^x and 3^x at the point where they cross the y-axis.

Coming up in the next section: more about logarithms. Stay tuned!

7.3: COMPOUND INTEREST

Too much of a good thing is wonderful.

—Mae West

In this project, we investigate compound interest and the effect of frequent compounding. Much to our surprise, we encounter once again that amazing number, e.

1. Suppose you put $1000 into an investment that pays 8% compounded annually. This means that, at the end of the year, 8% of the current balance is added to your account. How much will you have in the account after one year? After ten years? After t years?

2. Suppose, instead, that the 8% interest is compounded quarterly. This means that, at the end of every three months, 2% (that is, one-fourth of 8%) of the current balance is added to the account. How much will you have after one year? After ten years? After t years?

3. Your calculations in (1) and (2) should convince you that quarterly compounding is financially better for you than annual compounding. Investigate the effect of *monthly* compounding for one year (one-twelfth of 8% of the current balance will be added each month), and compare your results with the previous ones.

4. Is there any limit to the benefits of frequent compounding? Investigate the effects of weekly compounding (divide the interest rate by 52 and do 52 compoundings), daily compounding (use 365), and hourly compounding (use 8760 hours in a year). Suppose the interest were compounded every minute? Would you have more money at the end of the first year than you would with hourly compounding? Record the results of (1) through (4) in the table. (Don't do "continuously" yet.)

Frequency of compounding	Balance after one year
Once a year	
Quarterly	
Monthly	
Weekly	
Daily	
Hourly	
Each minute	
Continuously	

5. Some banks offer what they call *continuous compounding.* This does not mean that they actually work nonstop all year calculating the interest on your account, but that they use a mathematical formula like this one

$$P_0 \cdot e^{rt},$$

where P_0 is the original amount invested, r is the annual interest rate, and t is the number of years. Evaluate this formula for your investment, using 0.08 for r and one year for the time. Enter the result in the table and compare it with the other balances.

6. The exponential functions you used in (1) through (4) all had the same form:

$$P_0\left(1 + \frac{0.08}{n}\right)^n$$

where n was the number of compoundings per year. Based on your investigations, what is the value of the expression

$$\left(1 + \frac{0.08}{n}\right)^n$$

as n becomes large without bound?

7. Use a calculator to approximate $e^{0.08}$ and compare it with your result in (6).

8. Investigate the function

$$y = \left(1 + \frac{1}{n}\right)^n$$

(this would be the function to use if the bank were paying 100% interest) for very large values of n. Its graph should appear to level off, approaching a horizontal asymptote. Estimate the value of that asymptote. Then overlay the graph of $y = e$, and describe what you see.

7.4: PYRAMID POWER

In the beginning of 1997, riots erupted in the tiny eastern European country of Albania, the poorest country in Europe. Albanian citizens were outraged over their financial losses, which were precipitated by the breakdown of several pyramid schemes. Perhaps half the families in Albania had poured their own savings, as well as contributions from family members working abroad, into these swindles. A pyramid scheme creates the illusion of success by offering a very large interest rate to investors. The money to pay the interest, however, comes not from the operations of a business but from the cash of new investors. Sooner or later, the racket must run out of new investors and collapse. Albania, just emerging from years of a repressive communist government, had no financial regulations to restrict the operation of these investment funds and thus to protect a financially naive population.

In this project, you will use what you know about exponential functions to model a representative pyramid scheme and see why it was destined for an early collapse.

One of the largest of the schemes promised gullible investors interest of 50% *per month.* That offer was particularly seductive because the banks were offering interest of only 8% *per year,* which didn't even keep up with inflation. Let's create a mathematical model of this pyramid scheme, run by a woman named Maksude Kademi, who was arrested for fraud when the pyramid tumbled.

As we usually do when we model a situation from the real world, we will simplify the problem. Assume that the company accepts investments in multiples of \$1000 only and that each \$1000 corresponds to a different investor. Let's take as the rate of return the 50% per month that Kademi promised her investors, and suppose she found 300 people to invest at the start. That is, the pyramid began with \$300,000. (The first investors did make out handsomely, and served as the best advertisers of the scheme.)

1. At the end of one month, how much interest is owed to the initial investors?

2. Assuming that Maksude Kademi wants to keep the original \$300,000 for herself, she must make this interest payment by convincing other people to invest enough to cover the interest owed to the original investors. At \$1000 per investor, how many *new* investors does the pyramid scheme need to rope in by the end of the first month?

3. At the end of the second month of operation, the pyramid scheme must come up with 50% interest for the original 300 investors, as well as for the investors obtained during the first month. How much money does she need for the interest payments at the end of the second month?

4. How many new investors does this amount require?

5. Continue this process for two more months, keeping in mind that she has to pay interest to *all* the investors, not merely the new ones. (For now, don't worry if you get a fractional number of investors. Remember that this is only an approximation to a very messy real-world situation.)
   (a) Total interest payment at the end of the third month:

   (b) Number of new investors needed to meet this payment:

   (c) Total interest payment at the end of the fourth month:

   (d) Number of new investors needed to meet this payment:

Tirana investor loses his shirt AFP/Babani

6. Your results for months 2, 3, and 4 indicate that you can obtain the number of new investors at the end of the kth month by multiplying the number of new investors in the $(k-1)^{\text{th}}$ month by the same number. What is that constant multiplier?

A characteristic of any exponential function $f(x) = C \cdot b^x$ is that an increase of one unit in the value of the independent variable x results in the multiplication of the dependent variable by the base b. Therefore, your answer to (6) gives you the base of the exponential function that models this pyramid process.

7. Using this property of exponential functions, write a model for this pyramid scheme. The output $f(n)$ should represent the number of new investors needed at the end of month n, where $n = 1, 2, \ldots$. (Notice that n starts at 1, making this function a bit different from other exponential models, because $f(0)$ is not part of the model. So, in this case, C does not give the initial value for the model.)

   (a) You have already computed the first four y-values corresponding to inputs of 1, 2, 3, and 4. Evaluate $f(1)$, $f(2)$, $f(3)$, and $f(4)$, using the function you wrote, to be sure that your model is correct.

   (b) Use your model to compute the number of new investors needed at the end of the twelfth month (that is, one year into the swindle).

   (c) How many new investors would be needed at the end of two years (24 months)?

This problem is a lot like the checkerboard problem that began the AIDS lab. In the lab preparation, you computed the number of pennies on just the 64th square of the board to be 2^{63}. The number you computed in (7c) is the number of new investors required during just a single month, the final month of the second year. To compute the total number of investors that the scheme would need to dupe over the two-year period, we would have to add all the numbers: $300 + 150 + 225 + \ldots$. (This is similar to finding the total number of pennies on the checkerboard.) The total number of investors needed over the first two years is more than five million.[1] Since the entire population of Albania is only a little more than three million, and since there were several competing pyramid schemes in operation simultaneously, you can see why the collapse was inevitable.

In the United States, pyramid schemes are illegal, but there are marketing arrangements, such as some distributorships for cosmetics, operating on a similar exponential principle, whereby each new distributor recruits friends and acquaintances, and is promised a percentage of their commissions. Unlike the Albanian pyramid companies, however, the U.S. companies actually have products, so the investors get shampoo as well as promises.

[1]The corresponding total for the pennies is over 18,000,000,000,000,000,000—eighteen quintillion. We warned you that it would be a huge number!

7.5: MAKING A *BIG* DIFFERENCE

In this project, you will discover a fundamental distinction between polynomial functions and exponential functions. If you did *Making a Difference* (Project 5.3), you learned that the nth differences of a polynomial function are constant. Now we will examine the first and second differences for some exponential functions, with surprising results.

1. Complete the following table. The columns labeled "first differences" contain the results of subtracting two consecutive table entries for the function, and the columns labeled "second differences" contain the results of subtracting two consecutive first differences. The first differences are actually *average rates of change* of the functions.

x	x^2	x^2 first differences	x^2 second differences	2^x	2^x first differences	2^x second differences
-4	16			$\frac{1}{16}$		
		-7			$\frac{1}{16}$	
-3	9		2	$\frac{1}{8}$		$\frac{1}{16}$
		-5			$\frac{1}{8}$	
-2	4			$\frac{1}{4}$		
-1						
0						
1						
2						
3						
4						

2. Examine the first and second differences for x^2 and for 2^x. Write a couple of sentences describing and comparing the patterns you observe.

3. Calculate the first and second differences for the function 3^x. You have already seen that both the first and the second differences for 2^x form exponential functions themselves. In fact, they reproduce the values for 2^x. The patterns for the differences of 3^x will not be so apparent; nevertheless, they also form exponential functions. Try to find a formula for the exponential function produced by the first differences of 3^x and for the exponential function produced by the second differences of 3^x. The base of both of these new functions will also be 3, but you'll need to find the appropriate multiplier for each.

4. Calculate first differences for 4^x. Can you figure out an exponential function that fits these values? The base could be 4, but it could also be 2, because 4 is a power of 2.

The evidence you have gathered above should persuade you (even though we haven't proved it) that, while successive differences for any polynomial function lead ultimately to a constant difference, successive differences for an exponential function simply lead to another exponential function. When you get to calculus and your instructor tells you that the derivative of an exponential function is another exponential function, remember that you heard it here first.

7.6: VAMOOSE

Project 1.3, *A Big Moosetake,* reports an alarming increase in automobile-moose collisions in northern New England, attributable to the rapid increase in the number of moose in the region.

1. Suppose the moose population were to grow at a rate of 10% each year. At the start of 1991, there were an estimated 4000 moose in New Hampshire. How many *additional* moose would there be one year later? What would be the moose population at the start of 1992? At the start of 1993?

2. Instead of calculating the moose population year by year, we'll write a formula that estimates the number of moose in a given year t. Let t stand for the number of years since 1991. Given a 10% yearly growth rate, the number of moose in subsequent years can be modeled by an exponential function in standard form:

   $$f(t) = C \cdot b^t$$

   Write the formula for this model, specifying values for the constants C and b.

3. The standard-form exponential model you just wrote has an equivalent base-e form,

   $$f(t) = C \cdot e^{kt}$$

   Determine the values of C and k, and write the base-e form of the model. Then verify that, for both versions, $f(1)$ agrees with the number of moose you calculated for 1992, and $f(2)$ agrees with the number you calculated for 1993.

4. Approximately how many moose would you expect in New Hampshire in the year 2001, if this growth process were to continue unchanged for ten years?

5. Now, imagine that changing conditions (an increase in the deer population, perhaps, or loss of forest through development) cause the moose population to *decrease* at a rate of 15% a year after the year 2001. Thus, for $t > 10$, the number of moose would be approximated by a different exponential function. How many moose would we expect in 2002? (Use your result from (4) for the calculation.)

6. Now you'll write this new function in standard $C \cdot b^t$ form. There's something tricky here, because $t = 0$ is still supposed to stand for 1991, but we want this particular function to begin in 2001. A very easy way to get over the hurdle is to perform a *shift* of ten years to the right. Write the function in standard form and then modify the exponent to incorporate the shift. Verify that your function gives the correct results for 2001 and 2002 (years 10 and 11).

7. Write the new function in base-e form. Don't forget the 10-year shift.

8. Sketch the graph of the piecewise exponential function that models the changing number of moose in New Hampshire over the thirty year period from 1991 to 2021. Provide scales and label the axes.

9. Write the algebraic formula for this piecewise exponential model.

10. Assuming that this rate of decrease continues, predict the year by which the moose population in this region will be practically nonexistent. (You'll have to decide for yourself what constitutes being "practically nonexistent.")

11. In the previous questions, we assumed that the moose population would increase or decrease by a certain percentage each year. Suppose, instead, that the moose population in New Hampshire were to increase at the constant rate of 100 moose per year from 1991 to 2001 and that after 2001 the population were to decrease at a constant rate of 150 moose per year. Write a formula that models the number of moose in New Hampshire, with respect to time, for the years from 1991 until the moose have disappeared from the region. Your formula should be written as a piecewise linear function. Be sure that it gives the correct results for 1991 and 2001.

12. Sketch the graph of this function. Provide scales and labels.

13. According to this model, in what year will the moose disappear from New Hampshire?

7.7: LIMITS TO GROWTH — A Logistic Model

This project is a follow-up to Lab 7A

As you saw in the AIDS lab, an exponential function can model a real-world situation over only a limited time period. After a while, we begin to run out of people, or space, or money, or whatever it is whose growth we are studying, and the growth has to slow down; that is, the *rate* of growth must decrease.

Notice how the rate of growth of an exponential function like e^x or $4.244(1.825)^x$ is itself increasing over time: the functions *increase at an increasing rate.* Not only are they going higher, they're getting steeper as well.

In this project, you will investigate a function that starts out looking exponential but before long begins to feel the effects of its growth limits.

1. Make a freehand sketch of a graph to model a growth process in which there's a limit to the growth. Let A_0 represent the initial amount and suppose that the limit is given by some constant value A_{max}. (Assume A_{max} is greater than A_0.) Think a bit before drawing the graph. The growth limit will begin to exert its influence on the rate of growth as the value of A approaches the limit. The graph shouldn't get progressively steeper and then suddenly slam into the "ceiling."

2. Describe, qualitatively, what happens over time to the *growth rate* of $A(x)$. Refer to your sketch, which you might want to annotate.

In the AIDS lab, you used the exponential-growth function $f(x) = 4.244(1.825)^x$ to model AIDS in a population of women. You observed that the curve was too high for the data for the final two years and looked as if it would not work very well for future years. We mentioned that a more sophisticated function might work better as a model for these data. Here it is:

$$A(x) = \frac{1895.175}{1 + 184.623e^{-0.527x}}$$

where x continues to represent time in years since 1982.

3. Use a grapher to examine this function, which is called a **logistic growth** model. (*Caution:* enclose the entire denominator, as well as the exponent, in parentheses.) Try several different viewing windows. Does it have a shape similar to your free-hand sketch?

4. How many cases does it give for 1991 and 1992 (years 9 and 10)? Compare these values with the actual figures for those years, which were 726 and 972, respectively. Compare them with the values given by the exponential model $f(x) = 4.244(1.825)^x$ for those years.

5. Compare the two functions over the long run. What values do they have after 15 years? After 20 years?

6. What is the approximate rate of growth, in cases per year, of $A(x)$ when $x = 15$? What is the approximate rate of growth, in cases per year, of the exponential model $f(x)$ when $x = 15$? (Here you need to determine the *slopes* of the functions at 15, rather than their values at 15.)

7. Does the logistic model appear to have an upper limit? If so, what is it?

8. On the graph $A(x)$, locate the approximate point at which the function no longer increases at an *increasing* rate and begins to increase at a *decreasing* rate. Mathematicians call that point an **inflection point**. In approximately what year does the inflection occur?

9. Without doing any calculations, give the approximate rate of growth of $A(x)$ for the year 2106 ($x = 25$). How do you know?

10. Write a brief paragraph comparing and contrasting the logistic growth model and the exponential growth model, emphasizing the way in which their growth rates behave. In your description, be sure to distinguish between the functions themselves and their rates of growth.

Lab 8: Earthquakes

Shake, rattle, and roll

PREPARATION

> The earthquake, which struck at 12:30 A.M. Thursday, was measured at Teheran University at 7.3 on the Richter scale, indicating a severe quake. It combined with more than 100 aftershocks to cause huge landslides that blocked roads and relief convoys, knocked out electricity and water supplies in many areas, destroyed crops and left many villages isolated and thousands of people buried in the rubble of their homes. [1]
>
> —"Toll in Iranian Quake Is Put at 29,500,"
> *The New York Times*, June 23, 1990

The map shows the region described in the *New York Times* article. It pinpoints the earthquake's epicenter and outlines the regions hit hardest by the natural disaster.

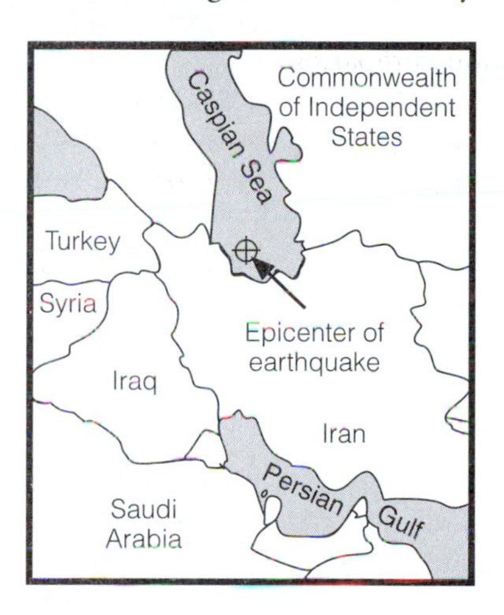

Earthquakes are dramatic events that attract a great deal of media attention. Coverage of earthquake disasters includes the death toll, reports of destruction, and estimates of economic impact, as well as specific information on the earthquake. Location of the earthquake's *epicenter* (the spot on the earth's surface directly above the region where ground movement first begins) and its size as measured on the Richter scale are two pieces of information contained in nearly every report.

Have you ever wondered about the technical details contained in such news reports? For example, how did scientists determine that the epicenter for the 1990 Iranian earthquake was in the Caspian Sea? How much stronger in terms of ground movement was this quake than the 1989 earthquake in California, which measured 6.9 on the Richter scale?

Modern seismographs located around the world (hundreds are in California alone) record a number of different types of earth movements. Two types of motions that can be detected are P waves and S waves, both of which radiate out in rings from the epicenter.

P waves cause back-and-forth movements, whereas S waves result in side-to-side motion. These two types of waves travel at different rates, the P waves arriving first, followed by the S waves. Just as the time delay between thunder and lightning can be used to estimate distance from a storm, the time delay between the arrival of the P waves and the S waves can be used to determine the distance from the seismograph to the epicenter.

A seismograph, then, can measure both the strength of a quake and its distance from the seismographic station, but it cannot by itself determine the direction from which the P waves and the S waves are arriving. To know that, geologists need information from more than one station. Here's a geometric method for finding the epicenter of an earthquake.

The map on the following page has five seismographic locations. Assume that an earthquake generates the following data (the distances are in arbitrary units):

Location	Distance from epicenter
1	2
2	6
3	7
4	6
5	7

Using a compass (the circle-drawing kind), locate the epicenter of this hypothetical earthquake on the map opposite. (If you do not have a compass, you can make one from a piece of string and a pencil. Tie the string to the pencil. From the pencil, mark off a length of string matching the distance from a particular location. Hold the mark at that location and pull the pencil away until the string is taut. Put the point of the pencil on the paper, keep the string taut, and draw an arc. Thus, every point on the arc is the desired distance from the location.) What was the minimum number of seismographic locations needed to pinpoint the epicenter of the earthquake?

All earthquakes are compared to what is called a zero-level earthquake—that is, one that would produce readings of 0.001 millimeter on a seismograph located 100 miles from the epicenter. (Thus, an earthquake with a seismograph reading of 0.02 is twenty times as strong as the zero-level earthquake, because $\frac{0.02}{0.001} = 20$.) The following is a hypothetical list of seismographic readings. (Assume that the readings have been adjusted for distance, so that the numbers can be compared with one another.)

$$0.003, \quad 12.7, \quad 0.02, \quad 1999, \quad 0.6, \quad 332$$

Try to plot these six readings on a real number line. What difficulty do you encounter?

The magnitude, $M(x)$, of an earthquake as measured on the Richter scale is computed by taking the base-10 logarithm of the ratio of the seismograph reading to the zero-level reading, 0.001. You can express this relationship algebraically as follows,

$$M(x) = \log\left(\frac{x}{0.001}\right)$$

where x is the seismograph reading in millimeters (adjusted for distance from the epicenter).

Compute the Richter scale magnitude of each hypothetical earthquake from the list of seismograph readings given above. Now, plot these Richter scale magnitudes on a real number line. Why are these numbers easier to plot than the first set?

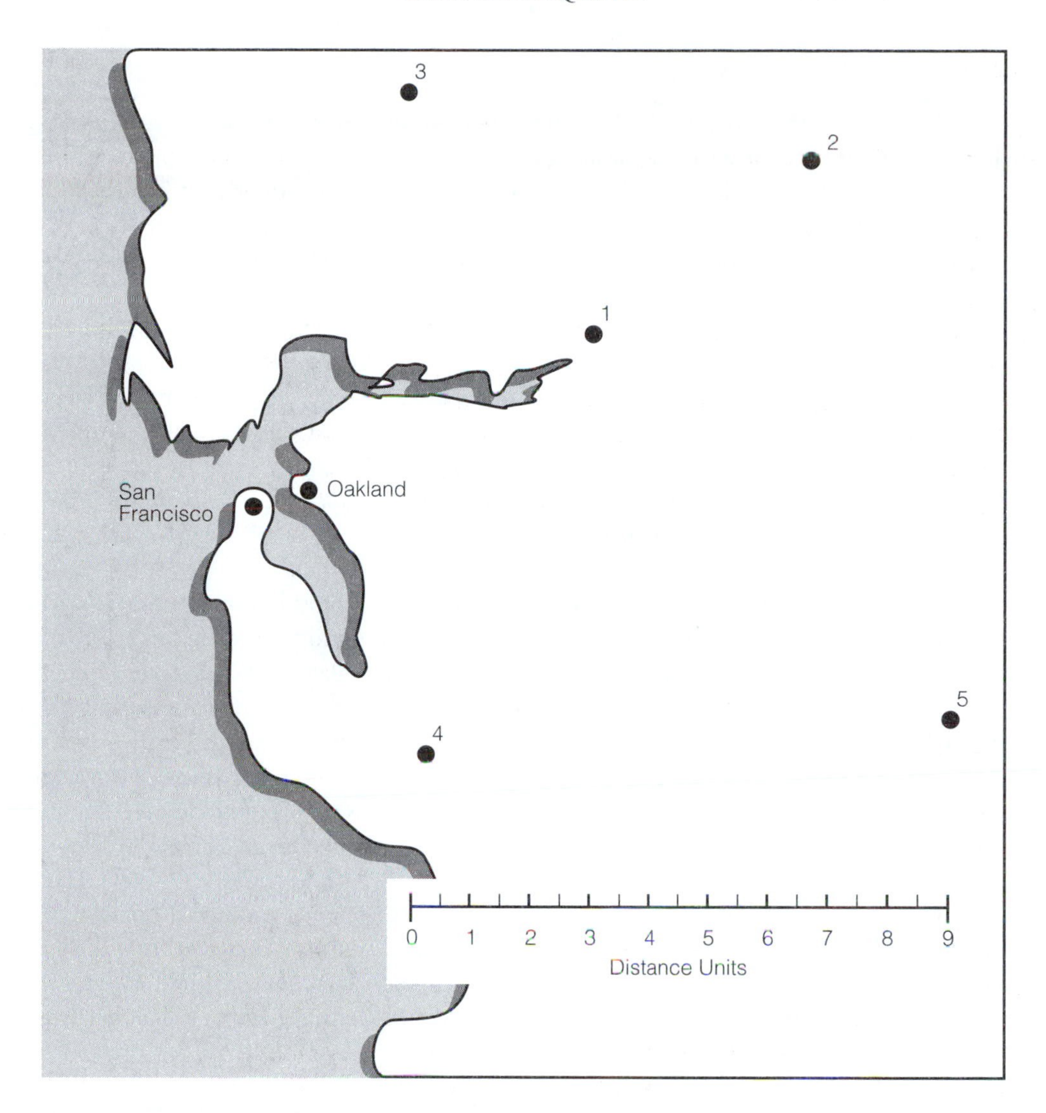
3
2
1
San
Francisco
Oakland
5
4
0
1
2
3
4
5
6
7
8
9
Distance Units

THE EARTHQUAKES LAB

Begin this lab by comparing with others in your group the work you did in the preparation section. Did you all center the earthquake at the same location? How many seismographic locations did you need? Discuss the difficulties you encountered in plotting the "raw" seismograph readings. How did your conversion of these readings into their Richter scale equivalents make the data easier to represent on a number line?

Because Richter magnitude is not well understood by the general public, reporters sometimes explain that an earthquake measuring 6 on the Richter scale is ten times as powerful as an earthquake measuring 5 on the Richter scale, and that an earthquake that measures 7 is 100 times as powerful as an earthquake that measures 5. Examine the algebraic formula for $M(x)$ and explain how the factors of 10 and 100 were determined.

If two earthquakes have Richter numbers of 7 and 3, respectively, how many times as strong as the second one is the first one?

Graph the function $M(x) = \log\left(\frac{x}{0.001}\right)$. The screen should show a curve that lies entirely on the positive side of the y-axis. Why does the function not exist for $x \leq 0$?

What about the output values? Can $M(x)$ itself ever be negative? If so, for what values of x does this happen? How large, in theory, can the function $M(x)$ grow? Answering these questions will demand some serious searching, with many adjustments to the viewing window. It is difficult to obtain a complete view of the graph of $M(x)$ without taking several different "snapshots." Don't rule out using algebra as a tool to help you here.

As an abstract function, what is the domain of $M(x)$? What is its range?

The graph shows a function that *increases* for all values of x. The exponential functions you saw in the AIDS lab were increasing functions, also, but they increased in a different manner, bending upward as they rose. We say that such exponential functions *increase at an increasing rate.* What comparable statement could we make about this logarithmic function?

Now, you will decide upon a reasonable domain and range for the mathematical model. Do you think that the portion of the graph where $M(x)$ is negative is relevant to the function M used as a mathematical model for earthquake strength? Consider the fact that the smallest earthquake that humans would be likely to notice measures approximately 2.5 on the Richter scale. Very sensitive seismographs, however, can detect local tremors whose magnitudes are -2.0 or even lower.

The table listing some major earthquakes of this century can give us an idea for a reasonable upper bound for the model.

What portion of the graph of $M(x)$ do you think actually represents the model? Give the domain and the range for the model by setting reasonable bounds for Richter scale readings and corresponding bounds for the seismograph readings.

Some Recent Major Earthquakes

Year	Place	Richter number	Deaths	Year	Place	Richter number	Deaths
1997	Iran	7.1	2,400	1978	Iran	7.7	25,000
1994	California	6.6	61	1977	Romania	7.5	1,541
1993	India	6.5	21,000	1976	Turkey	7.9	4,000
1992	Indonesia	7.5	2,500	1976	Philippines	7.8	8,000
1992	California	7.4	63	1976	China	7.8	200,000*
1990	Iran	7.3	29,000	1976	Guatemala	7.5	22,778
1990	Philippines	7.7	440	1974	Pakistan	6.3	5,200
1989	California	6.9	270	1970	Peru	7.7	66,794
1988	Armenia	6.9	25,000	1970	Turkey	7.4	1,086
1985	Mexico	8.1	9,500	1968	Iran	7.4	12,000
1983	Turkey	7.1	1,300	1966	Turkey	6.9	2,520
1980	Italy	7.2	4,800	1964	Alaska	8.4	131
1980	Algeria	7.3	4,500	1962	Iran	7.1	12,230
1979	Colombia-			1960	Chile	8.3	5,000
	Ecuador	7.9	800	1960	Morocco	5.8	12,000

*Official estimate; may have been considerably higher.

Seismologists do not always agree on the precise magnitude of a given earthquake. While officials at Teheran University measured the Iranian earthquake to be 7.3 on the Richter scale, the United States Geological Survey at Golden, Colorado, reported a measurement of 7.7. Use your graph of $M(x)$ to compare the higher reading to the lower reading. How many times as strong as the estimate from Teheran was the estimate from Colorado?

The strongest aftershock of the Iranian earthquake measured 6.5 on the Richter scale. How many times as strong as this aftershock was the 1989 earthquake in California?

Stop and look carefully at your results for the two comparisons you just performed. Both times, you should get the same answer. Can you find a pattern relating the Richter numbers you were asked to compare? A solid understanding of the mathematics involved will not only help you to appreciate how a logarithmic scale operates but will point the way to a very fast method for comparing the strengths of any two earthquakes from their Richter magnitudes. Don't move on until you understand this shortcut.

In addition to the Richter scale magnitude, the table contains the estimated number of earthquake-related deaths. Does there appear to be any correlation between the number of deaths and the Richter magnitude?

Two articles appeared on the same page of *The Boston Globe* on August 16, 1991. "Nuclear Weapon Tested in Nevada" reported the detonation of a nuclear weapon 1600 feet underground. The test caused the ground to sway, registering 4.4 on the Richter scale. "Quake Put at 3.0 Hits Central Pa." reported on a light earthquake with epicenter 10 miles from State College. The earthquake that measured 3.0 on the Richter scale was the strongest to hit this area since 1944. Please comment on these reports. You might use your newfound ability to compare two quakes in your comments.

Earthquake measurement is only one example of a logarithmic scale. Measuring the loudness of sound in decibels and the acidity of rain on the pH scale are two other examples. In Lab 10, you will meet the Bordeaux Equation, in which the output Q is a logarithm (that is, an exponent), and needs to be put into the form e^Q to yield a meaningful number. In that lab, you will see bunched results (Q-values) become nicely spread out when they are converted to e^Q-values. In this lab, you saw input data (ground movement measurements) that were too spread out but which yielded meaningful numbers (Richter magnitudes) after logarithmic scaling.

THE LAB REPORT

Describe how you used geometry to locate the epicenter of an earthquake and how you determined the minimum number of seismographic stations needed to pinpoint its location. Explain to someone not in this course how the Richter magnitude is determined from the seismograph reading and how it describes earthquake strength. Be sure to include reasons *why* we would want to use a logarithmic scale in this case. Support your reasons with specific examples. Show how to use Richter magnitudes to compare the strengths of two earthquakes. Illustrate your discussion with one or more graphs of the Richter function $M(x)$ and indicate the portion of the graph that gives reasonable bounds for Richter earthquake magnitudes.

Finally, report any other observations your group made. Did there appear to be a relationship between the magnitude of an earthquake and the number of resultant deaths? (Include specific examples and speculate as to other factors, besides Richter magnitude, that might contribute to the differences in death tolls.) What were your reactions to the two *Boston Globe* items from August 1991?

Section 8—Projects and Explorations

8.1: TRAVELOG—Exploring Logarithmic Functions

The Natural Log Function

1. Draw a graph of the natural (base e) log function, $f(x) = \ln x$.

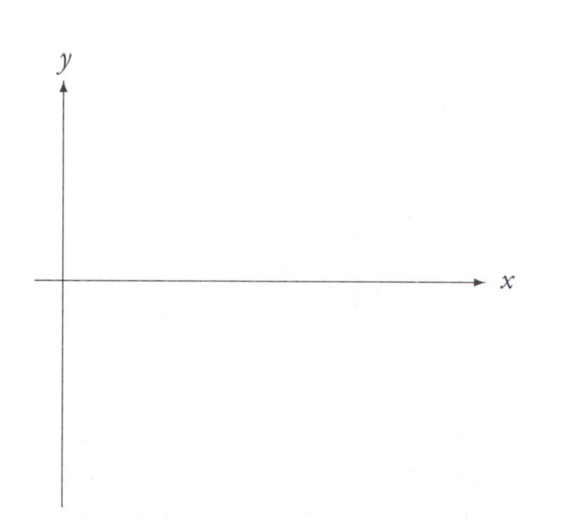

Label the x-intercept of the graph. For what values of x is $\ln(x)$ negative? Positive?

Is there a y-intercept?

2. Notice that, as x approaches 0 from the right side, the output values fall rapidly. We say that the y-axis is a vertical asymptote for $\ln(x)$. To get a feel for how rapidly the natural log of x is falling as x values are chosen closer and closer to zero, fill in the following table.

x	$\ln x$
0.01	
0.001	
0.0001	
0.00001	

Recalling that a logarithm is an exponent, explain why we should expect this set of inputs to have negative logs, and why the logs should become more and more negative as the numbers get closer to 0.

3. Let's look at how this function behaves if we back away and look at the graph from a distance. Reset the width of the window to $0 \le x \le 100$. (You may have to adjust the y-values as well to keep the curve on the screen.) Does the graph level off or does it keep rising? Repeat, using a viewing window with $0 \le x \le 1000$.

4. Find the value of x for which $f(x) = 6$. (You might want to overlay the graph of the constant function $g(x) = 6$ and find the x-value of the point of intersection.) Find the values of x for which $f(x) = 7$ and for which $f(x) = 8$. How many x-units were there between $f(x) = 6$ and $f(x) = 7$? Between $f(x) = 7$ and $f(x) = 8$?

5. Does $f(x)$ ever get bigger than 20? Than 100? (Is your grapher up to the task of answering this last question?) Is there a ceiling on the values for $\ln x$? If so, what would the value of the ceiling be?

6. Now, with confidence, write the domain and the range of the natural log function.
   (a) The domain of $\ln(x)$:

   (b) The range of $\ln(x)$:

Log Functions With Other Bases

7. Compare the graphs of $\log_{10}(x)$ and $\ln(x)$. In what ways are they the same? How do they differ?

8. Few graphers will handle log functions with other bases without a little help from us. To view the graph of $\log_2(x)$, say, we need to convert the expression into a form that the grapher will understand. Let $y = \log_2(x)$. Then, rewrite that equation in exponential form: $2^y = x$. Take the natural logarithm of both sides of this new equation and solve for y. You have just rewritten $\log_2(x)$ in a form that the grapher can understand. Graph this function, along with $\log_{10}(x)$ and $\ln(x)$, and compare all three.

9. The technique you just used to rewrite $\log_2(x)$ works equally well for logarithms whose base is any positive number (other than 1). Use it to graph $\log_5(x)$, and compare its graph to the previous three. How does the size of the base affect the graph of a logarithmic function?

8.2: ANALOG—Algebraic Equivalences with Logs

This project is a follow-up to Lab 8

1. In the Earthquakes lab, the relationship between the seismograph readings x and Richter scale magnitude $M(x)$ was given by

$$M(x) = \log\left(\frac{x}{0.001}\right)$$

Sometimes the Richter function is presented as

$$M(x) = \log(x) + 3$$

Graph the two expressions for $M(x)$. How do the two graphs compare? Show algebraically why you should expect such a result.

Because the functions e^x and $\ln(x)$ are inverses of each other, we see some interesting and maybe surprising results when we combine them.

2. Graph the function $\ln(e^x)$. What do you see? Give a simpler expression for this function.

3. Clear the screen and graph the function $e^{\ln(x)}$. Describe what you see. It should be much the same as the graph of $\ln(e^x)$, but with one important difference.[1] What is that difference, and why should we expect it?

[1]Certain graphers, such as MAPLE, DERIVE, and the TI-92, will show the same graph for (2) and for (3), because they're using complex (nonreal) numbers. If you see identical graphs, ask yourself which *real* numbers are not in the domain of $e^{\ln(x)}$.

4. Compare the graphs of $e^{-\ln x}$ and $\frac{1}{x}$. How are they alike? In what way are they different? Explain the similarity by simplifying the first expression using rules for exponents and logarithms. Explain the difference by recalling the domain of the natural log function.

5. Graph the function $f(x) = 2\ln(x)$ and compare it to $\ln(x)$. Explain why you should see a vertical stretch.

6. Now consider the function $g(x) = \ln(x^2)$. The rules for logs say that $\log(N^p) = p\log(N)$, so we are entitled to expect the graph of $g(x)$ to resemble the graph of $f(x)$ from (5). Not so fast, though! First, state the domain of $f(x)$. (Be careful!) What type of symmetry should we expect in the graph of $g(x)$?

7. Now view the graph of $\ln(x^2)$ to see if it is what you predicted.

8. Instead, let $f(x) = \ln x^3$ and $g(x) = 3\ln x$. Is $f(x)$ equivalent to $g(x)$? Provide both graphical and algebraic support for your answer.

8.3: I HEAR THE TRAIN A-COMIN'

This project is a follow-up to Lab 8

The human ear is sensitive to a wide range of sound intensities. Sound becomes audible after it reaches an intensity of 10^{-16} watt/cm^2 and becomes painful to the ear at about 10^{-4} watt/cm^2 (the threshold of pain).

1. How many times as great as the sound intensity for the faintest audible sound is the sound intensity at the threshold of pain?

In the Earthquakes lab, you saw that measurements of earthquake strength given by seismograph readings, x, became more manageable when converted to a logarithmic scale, the Richter scale. Specifically, this relationship was given by the function

$$M(x) = \log\left(\frac{x}{0.001}\right)$$

Similarly, converting sound intensity into a base-10 logarithmic scale produces numbers that are easier to use. The sound intensity level $\beta(x)$, measured in decibels, is related to the sound intensity x, measured in watt/cm^2, by the formula

$$\beta(x) = 10\log\left(\frac{x}{x_o}\right)$$

where $x_o = 10^{-16}$ watt/cm^2 (a reference intensity corresponding to the faintest sound that the average person would be able to hear).

2. What is the sound intensity level in decibels for the faintest sound that can be heard by the average person? What is the decibel level for the threshold of pain?

3. The sound intensity level of a quiet whisper is about 20 decibels, while that of an an ordinary conversation is about 65 decibels. Compute the sound intensities, measured in watt/cm^2, for the quiet whisper and ordinary conversation. Compare the two numbers you just computed: how many times greater than the smaller is the larger?

4. A difference in sound intensity level of one decibel is generally taken to be the smallest increase in volume that the average human ear can detect. How much of an increase in watt/cm^2 will be required to raise the sound level of a 20-decibel whisper by one decibel? How much of an increase will be required to raise the level of a 65-decibel conversation by one decibel?

5. The sound of a train passing over a trestle measures about 90 decibels. Suppose a particular train caused a noise of 91 decibels. Would the increase in watt/cm^2 corresponding to a change from 90 to 91 decibels be greater or less than the increase required to go from 65 to 66 decibels? Explain.

6. Examine the shape of the graph of $\beta(x)$. As in the Earthquakes lab, you will need to use several different intervals, ones that are reasonable in this context, in order to get a complete picture of a logarithmic curve. What happens to the graph as the sound intensity x gets very close to zero? How does the basic shape of this graph compare to the graph of $M(x)$ in the Earthquakes lab? Comment on the similarities between the formulas for $M(x)$ and $\beta(x)$.

Lab 9: Daylight and SAD

You Are My Sunshine

PREPARATION

Before coming to the lab, you should make yourself familiar with the graph of the sine function. You should know the meaning of the terms **amplitude**, **period**, and **phase shift**. Sketch the following graphs. Show scales on the axes.

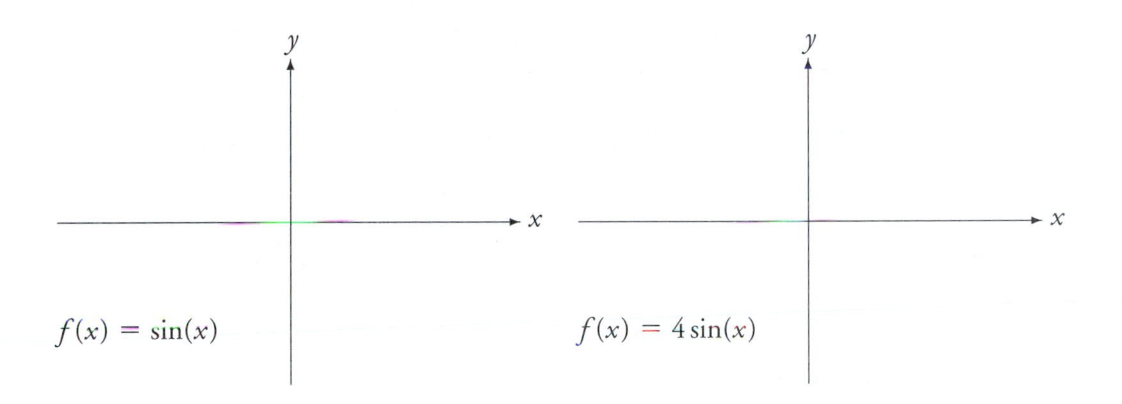

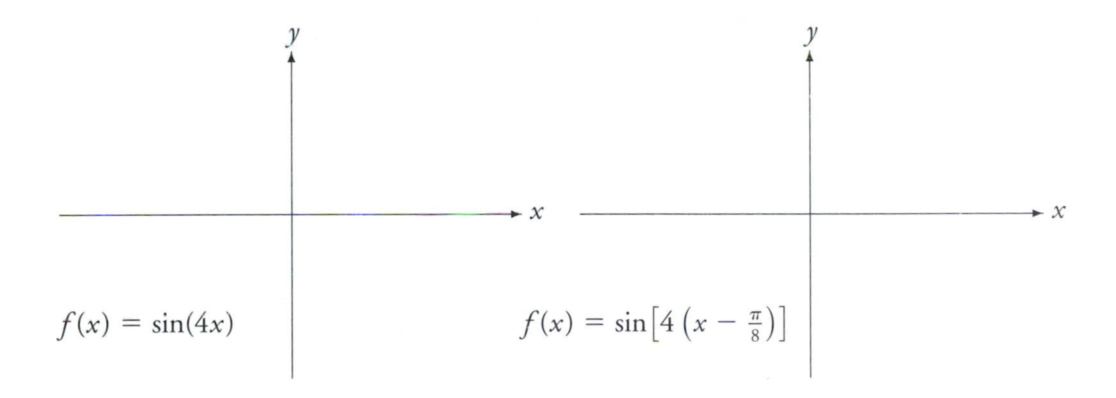

If your grapher can't plot individual points, you should also prepare, on graph paper, two graphs of daylight data from the tables in the lab pages: one for Boston and one for Reykjavik. To be able to see the annual pattern more clearly, plot two years' worth of points. **Bring the graphs with you to the lab.**

If the grapher does have point-plotting capability, plan ahead how you will enter the data so that you can see a two-year curve, assuming 365 days in the year.

THE (SAD) LAB

Over the course of a year, the length of the day—that is, the number of hours of daylight, calculated by subtracting the time of sunrise from the time of sunset—changes every day. Here is a table giving the length of day, rounded off to the nearest tenth of an hour, for Boston, latitude 42° N.

Date	Day	Hours of daylight	Date	Day	Hours of daylight
1/2	2	9.2	7/5	186	15.2
1/10	10	9.3	7/13	194	15.0
1/18	18	9.6	7/21	202	14.8
1/26	26	9.8	7/29	210	14.5
2/3	34	10.1	8/6	218	14.3
2/11	42	10.4	8/14	226	13.9
2/19	50	10.8	8/22	234	13.6
2/27	58	11.1	8/30	242	13.3
3/7	66	11.5	9/7	250	12.9
3/15	74	11.9	9/15	258	12.5
3/23	82	12.3	9/23	266	12.1
3/31	90	12.7	10/1	274	11.8
4/8	98	13.0	10/9	282	11.4
4/16	106	13.4	10/17	290	11.0
4/24	114	13.8	10/25	298	10.7
5/2	122	14.1	11/2	306	10.3
5/10	130	14.4	11/10	314	10.0
5/18	138	14.8	11/18	322	9.6
5/26	146	14.9	11/26	330	9.5
6/3	154	15.1	12/4	338	9.3
6/11	162	15.2	12/12	346	9.2
6/19	170	15.3	12/20	354	9.1
6/27	178	15.2	12/28	362	9.2

Plot these points (day of year versus hours of daylight) on your grapher, if it has point-plotting capability. Otherwise, use the hand-plotted graph you prepared. Plot two years' worth of data so that you can see the pattern repeating itself.

The shape you see should look like a rough approximation of a sine wave. In fact, the graph can be approximated by a function of the form

$$f(x) = A\sin[B(x - C)] + D$$

Our challenge in this lab will be to determine values for the constants A, B, C, and D.

Let's first consider A, the **amplitude**, which determines the height of the wave. The constant A tells how far the wave will stray from its center line. How tall does the daylight wave need to be? Use the table to give you some ideas. Because we don't have data for every day of the year and the values we do have have been rounded off, we don't know exactly what the highest and lowest values are or on which days they occur. What might be reasonable numbers to use for the highest and lowest values, given the data that we have? Decide a value for A and check to see that it gives your sine wave the proper height.

Next, let's find B, which provides information on the **period** of the function. A standard sine function repeats its pattern every 2π units; that is, it has a period of 2π. What is the period of the daylight function? (After how many days does the length-of-day pattern repeat itself?) We will need to stretch out the standard sine wave if we want it to match the daylight pattern. Use the desired period, along with the 2π, to determine the necessary stretching factor, B. Keep in mind that the B-value itself is not the period of the function, but the horizontal stretching factor that produces the correct period.

How are you doing? Graph what you have so far, $A\sin(Bx)$, using your values for A and B, and compare its graph with the daylight graph. You should see a curve that has the correct size and shape but is in the wrong place.

Skip over C for the moment and consider D, which gives the average value of the sine wave. You probably see that your graph is too low, oscillating about the x-axis instead of up where the data points are. How much higher should it be? How did you decide? Use D to give your graph the boost it needs.

Once you determine D, you know the horizontal axis of the sine wave (overlaying the line $y = D$ would be helpful) and you can work on finding C, which gives the **phase shift** of the function.

Start by determining visually how much of a horizontal shift (how many days) your graph needs, and in which direction. Remind yourselves how to write a horizontal shift into a function, and don't forget that you'll need to modify the independent variable *directly*, before you do anything else (such as multiplication) to it. In other words, pay attention to parentheses.

Try the completed model, $f(x) = A\sin[B(x - C)] + D$, using your values for A, B, C, and D. Its graph should hug the data points closely. If not, decide which constant needs to be adjusted. No matter how hard you try, though, you'll never achieve a *perfect* fit.

Remember that the data are approximations and so were the points you plotted. If you used a grapher to plot the points, you should notice that few of the points went exactly where you tried to place them because there are a limited number of pixels on the screen and the grapher does the best it can. Much of what you did in choosing the constants involved intelligent estimation. You found a function that models the data fairly well. There is more than one such function, and no function will be a perfect match.

One difficulty inherent in finding a model is that the given pattern is *not* a perfect sine wave, since the apparent path of the sun is slightly elliptical rather than perfectly circular. That's a problem we are not going to attempt to fix in this lab!

The seasons affect everyone's moods to some degree, but some people are so strongly affected by the amount of daylight that they experience severe depression during that part of the year when the hours of daylight are shortest. Many magazine articles have been written about this condition, which has been termed seasonal affective disorder, or SAD. A typical person with SAD feels depressed for two or three months, sometime between the end of October and late February. She (women are affected more often than men) may experience a lack of energy and a craving for carbohydrates, and she may respond by oversleeping, overeating, and withdrawing from society. An estimated 6% to 8% of the population of New England suffers from full-blown SAD.

Unlike the traditional treatments for other forms of depression, an effective therapy for SAD has the patient sit in front of bright lights every morning. If we assume that one hour of light therapy is equivalent to an hour of natural daylight, approximately how many hours of light therapy might a person with SAD require on a day in early January if she wanted to make up for the "missing" hours of natural daylight (compared to March 21, the first day of spring)?

The graph that follows represents results from a study[1] of SAD patients and a group selected at random from the New York City telephone book, in which they were asked to specify the months in which they felt best or worst. Each point shows the proportion of people feeling at their best or worst in a particular month. "Feeling worst" is counted as a negative value.

If the seasonal mood fluctuations of SAD patients could be approximated by a sine function, would the function be in phase with the length-of-day function or out of phase? Explain.

Notice that the effect is reversed during spring and summer: SAD patients may, in fact, feel better than the average person. They are full of energy and usually lose the extra weight they put on during the winter.

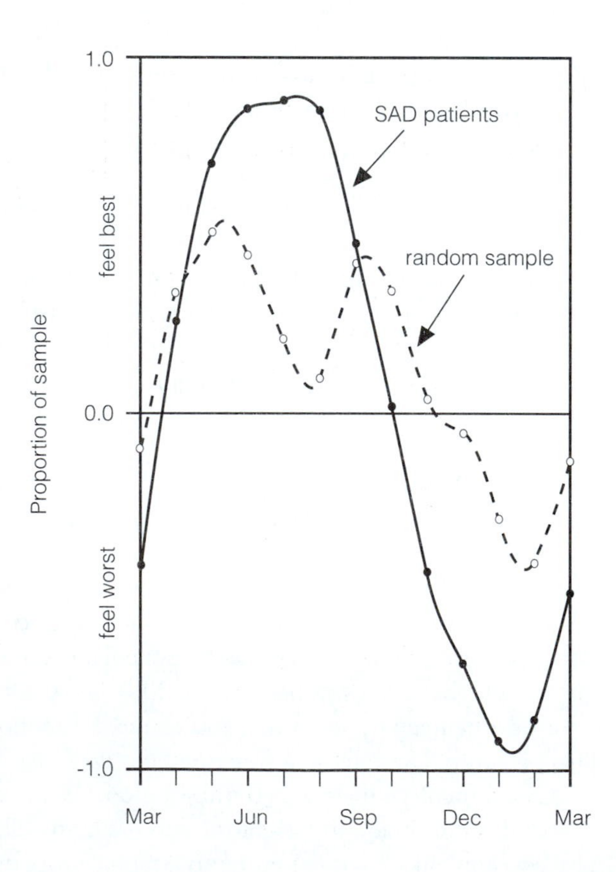

[1] Michael Terman, "On the Question of Mechanism in Phototherapy for Seasonal Affective Disorder: Considerations of Clinical Efficacy and Epidemiology," in *Journal of Biological Rhythms*, Vol. 3, No. 2, 155–172, 1988.

SAD appears to be even more prevalent farther north. Here are some data for Reykjavik, Iceland, latitude 64° N.

Date	Day	Hours of daylight	Date	Day	Hours of daylight
1/2	2	4.5	7/5	186	20.5
1/10	10	5.0	7/13	194	19.9
1/18	18	5.7	7/21	202	19.1
1/26	26	6.5	7/29	210	18.3
2/3	34	7.3	8/6	218	17.4
2/11	42	8.2	8/14	226	16.5
2/19	50	9.1	8/22	234	15.7
2/27	58	9.9	8/30	242	14.8
3/7	66	10.8	9/7	250	13.9
3/15	74	11.7	9/15	258	13.1
3/23	82	12.5	9/23	266	12.2
3/31	90	13.4	10/1	274	11.4
4/8	98	14.3	10/9	282	10.5
4/16	106	15.1	10/17	290	9.7
4/24	114	16.0	10/25	298	8.8
5/2	122	16.9	11/2	306	8.0
5/10	130	17.8	11/10	314	7.1
5/18	138	18.6	11/18	322	6.3
5/26	146	19.5	11/26	330	5.5
6/3	154	20.2	12/4	338	4.9
6/11	162	20.7	12/12	346	4.3
6/19	170	21.0	12/20	354	4.2
6/27	178	20.9	12/28	362	4.3

When you plot these points, you will see that they do not seem to lie nearly so close to a smooth sine curve as Boston's points did. The Reykjavik data are less precise and, therefore, you must not take any single point too seriously. Nevertheless, it's possible to find a sine function that fits the data, taken as a whole, fairly well.

Writing such a function, $A\sin[B(x - C)] + D$, won't involve much more work than you've already done. Think about what features of the graph remain substantially the same. Which letters (A, B, C, or D) control those features? What features of the Reykjavik graph makes it different from the Boston graph? Which constants control those features? Calculate new values for those constants, and try out your new mathematical model.

THE LAB REPORT

Your lab report should explain how you decided on the value of each constant in the Boston length-of-day function and how you modified that function to write one for Reykjavik. Mention any difficulties you encountered in attempting to fit a sine curve to the data. Explain what information these models might provide about seasonal affective disorder, including (among other things) the effect of latitude. Include any observations you made about the graph of seasonal mood fluctuations and its relationship to the daylight models. Illustrate your report with appropriate graphs.

On the light side

From 1948 until 1951 Japan practiced Daylight Savings Time. From April to September there was a summer time schedule of an additional hour of sunlight. This system was abolished because of the following reasons:

1. The sun set too late. An additional meal was required because the day was so long.
2. Longer hours for laborers.
3. Lack of sleep.

More information is available on request.

—*Japanese Embassy Information and Culture Center publication, reprinted in* The New York Times (1991)

Section 9—Projects and Explorations

9.1: COPYCATS—Exploring Sine and Cosine Functions

In this activity, you will investigate the effects of various modifications to the basic sine and cosine graphs, and will see them as examples of transformations that you have already learned—shifts, stretches, and so on.

1. On the interval $[0, 2\pi]$, draw a graph of $\sin(x)$ and of $\cos(x)$. You have drawn what we call a **fundamental wave** of each of those functions. Notice that the fundamental sine wave starts and ends on its center line, while the fundamental cosine wave starts and ends at its highest value.

   (a) $y = \sin(x)$

   (b) $y = \cos(x)$

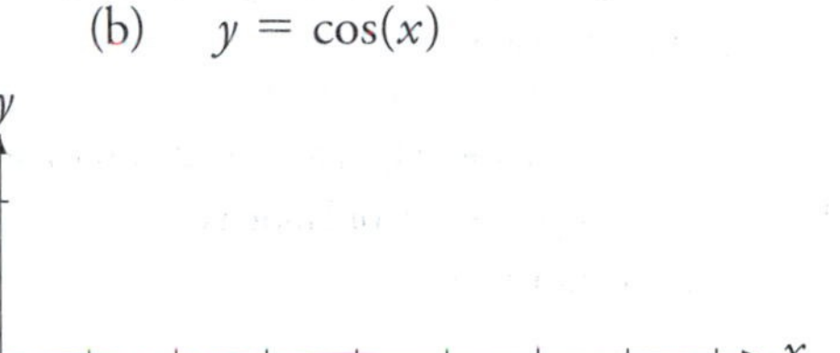

2. Experiment with the sine and cosine graphs by multiplying the functions by various constant values: $A\sin(x)$ and $A\cos(x)$. Use both positive and negative values for A, including some numbers between -1 and 1. Write a summary of your observations, interpreting the effects you see as the result of one of the transformations you have studied. The absolute value of the coefficient A is called the **amplitude** of the sine or cosine function. Does this seem to you to be an appropriate name for $|A|$?

3. The **period** of a sine or cosine function means the width of the smallest interval that contains one complete fundamental wave. What is the period of $\sin(x)$? Of $\cos(x)$?

(a) Speed up the wave: fit two sine waves into the interval $[0, 2\pi]$. What formula did you use to create the graph?

(b) Now fit three waves, four waves, five waves into that interval. Write the formula for each function you used.

(c) How would you slow the wave down, causing it to use twice as much horizontal space to complete its fundamental pattern? Write the function.

(d) Can you find a sine wave with a period of one unit? Give it a try. If you find yourself frustrated, move on and come back to this.

(e) Interpret the effects you see as the results of another transformation you have learned.

4. To complete (3), you needed to write functions of the form $\sin(Bx)$. The multiplier B does not have a mathematical name, but it is important to each of these functions because it determines the period. There is a connection between the value of B and the period of a sine or cosine function; now you'll discover their relationship.

(a) As B grows, does the period lengthen or shorten?

(b) As B shrinks toward 0, what happens to the period?

(c) If your answers to (a) and (b) are correct, they suggest that the value of B and the length of the period of $\sin(Bx)$ or $\cos(Bx)$ are *inversely proportional*—that is, that

$$\text{period} = \frac{k}{B}$$

where k is a constant that we will determine. Use the fact that the period is 2π when $B = 1$ to find k.

(d) Now write a formula for finding the period of $\sin(Bx)$ or $\cos(Bx)$ if you know B.

(e) Write another formula for finding B if you know the period.

5. Sine and cosine waves often come with a **phase shift**, a horizontal shift that causes them to begin their fundamental pattern somewhere other than in their accustomed place. You already know how to shift a graph laterally. Now you will apply that technique to sine and cosine graphs.

   (a) Shift the graph of $\cos(x)$ one unit to the left. What formula did you use?

   (b) Shift the graph of $\sin(2x)$ one unit to the left. Write the formula you used. Be sure that the graph shifted exactly one unit. If it didn't go far enough, check parentheses. Remember that, to cause a horizontal shift, we add a constant directly to the independent variable (before anything else happens to the variable).

   (c) Shift the graph of $\sin(x)$ to coincide with $\cos(x)$. What formula did you use?

   (d) Shift the graph of $\cos(x)$ so that it coincides with itself. Give the formula.

   (e) Shift the graph of $\sin(2x)$ so that it coincides with $\cos(2x)$.

6. Not all waves oscillate about the x-axis. Some are higher; others are below the axis.
   (a) Find a sine function that oscillates about the line $y = 12$.

   (b) Find a sine function with an amplitude of 3 whose top bumps the x-axis but does not cross it. (Mathematicians would say that the graph is *tangent* to the x-axis at its peaks.)

7. Now that you have investigated four different ways in which a sine or a cosine wave can be transformed, here are some practice problems. Predict the appearance of each wave in terms of its amplitude, its period, its phase shift, and its vertical shift. Then check your prediction by viewing the graph.
   (a) $-10\sin(2x) + 10$

   (b) $3\cos[0.25(x - 1)]$

   (c) $1 - \cos(\pi x)$

   (d) $\sin[2\pi(x + 0.1)]$

   (e) $\cos(3x + 1)$ (Be careful with the phase shift!)

9.2: SAD AND LATITUDE

This project is a follow-up to Lab 9

The incidence and severity of seasonal affective disorder seem to depend upon latitude. *Scientific American* (January 1989) reports that 24% of the population of Tromsø, Norway (latitude 69° N.), may suffer from midwinter insomnia, another manifestation of SAD. Tromsø is so far north (more than 200 miles above the Arctic Circle) that the people there do not even see the sun between November 20 and January 20. (They do, however, enjoy 24 hours a day of sun during the summer.)

1. Draw a sketch of what the length-of-day function for Tromsø might look like. Put a scale on each axis. Does your sketch show a sine wave? Explain.

2. You probably won't be able to write a precise mathematical formula for the function whose graph you have drawn. It is a piecewise function, some segments of which are constant. Which segments are constant? What are their constant values?

3. In the Southern Hemisphere, where the pattern of daylight is reversed, SAD reaches its peak during June and July. Put scales on the axes and sketch a graph of the length-of-day function for Wellington, New Zealand, or Puerto Montt, Chile, both of which have an approximate latitude of 42° S. (They are the same distance from the equator as Boston, but in the opposite direction.)

4. Show how to modify one of the mathematical models you derived in the SAD lab to obtain a formula for this graph.

9.3: DAY BY DAY—An Alternate Variety of SAD

This project is a follow-up to Lab 9

Some people with seasonal depression are more prone to feel symptoms in spring and fall than in winter. Their depression appears to be triggered not by a shortage of light, but rather by changing amounts of it. Psychologists wishing to help such patients might want to know the rate at which the amount of daylight is changing at various times of the year.

1. Using the length-of-day function you wrote for Boston, examine the graph and find the season of the year during which the days are lengthening most rapidly. (How can you tell from the graph which time of year that is?) Zoom in on that section, magnifying several times, until the portion you see resembles a straight line. Use the trace feature to find the coordinates of two points on that line and calculate its slope. With that information, determine how rapidly, in *hours per day*, the days are lengthening at that time of year. Now, convert your calculation into *minutes per day*, to the nearest whole minute. How many minutes per week is this? Would a person be likely to notice the difference from one week to the next?

2. Do the same steps for the part of the year when the days are shortening most rapidly. Compare the results with your previous ones.

3. Repeat the two sets of rate-of-change calculations, but this time use the length-of-day function for Reykjavik. Compare your results with those for Boston. The number of daylight hours changes from day to day much more rapidly in Reykjavik than it does in Boston; approximately how many times as rapidly does it change?

4. Examine the region of the graph (either graph) where the greatest number of daylight hours occurs. What is the approximate rate of change, in minutes per day, for that region? Explain how you can read this information from the graph without having to do any calculations.

5. A psychologist studying this particular form of seasonal depression asks what you're working on. Explain, in a brief paragraph, the correlation between the two sine functions and the variety of SAD described at the beginning of this project.

9.4: DOWN THE DRAIN

Engineers concerned with the public water supply need to know the daily water usage patterns for their community. A municipality typically has water storage tanks to provide a reserve for periods of high demand. When large quantities of water are being used, the amount of water stored goes down. During periods of light demand, the tanks refill from the municipal wells or reservoirs. Engineers monitor the water level in the storage tanks and can control the flow by opening or closing valves. They try to smooth out the flow so that a reserve is maintained.

The circular graph shows one week's readings from a device that measures the water level in a town's storage tank. The pen traces a path far from the center when the water in the tank is at a high level; the pen moves toward the center when the water is at a low level. Although there is some variation and some asymmetry, we can see a pattern here: the level is at its highest every morning and at its lowest every evening.

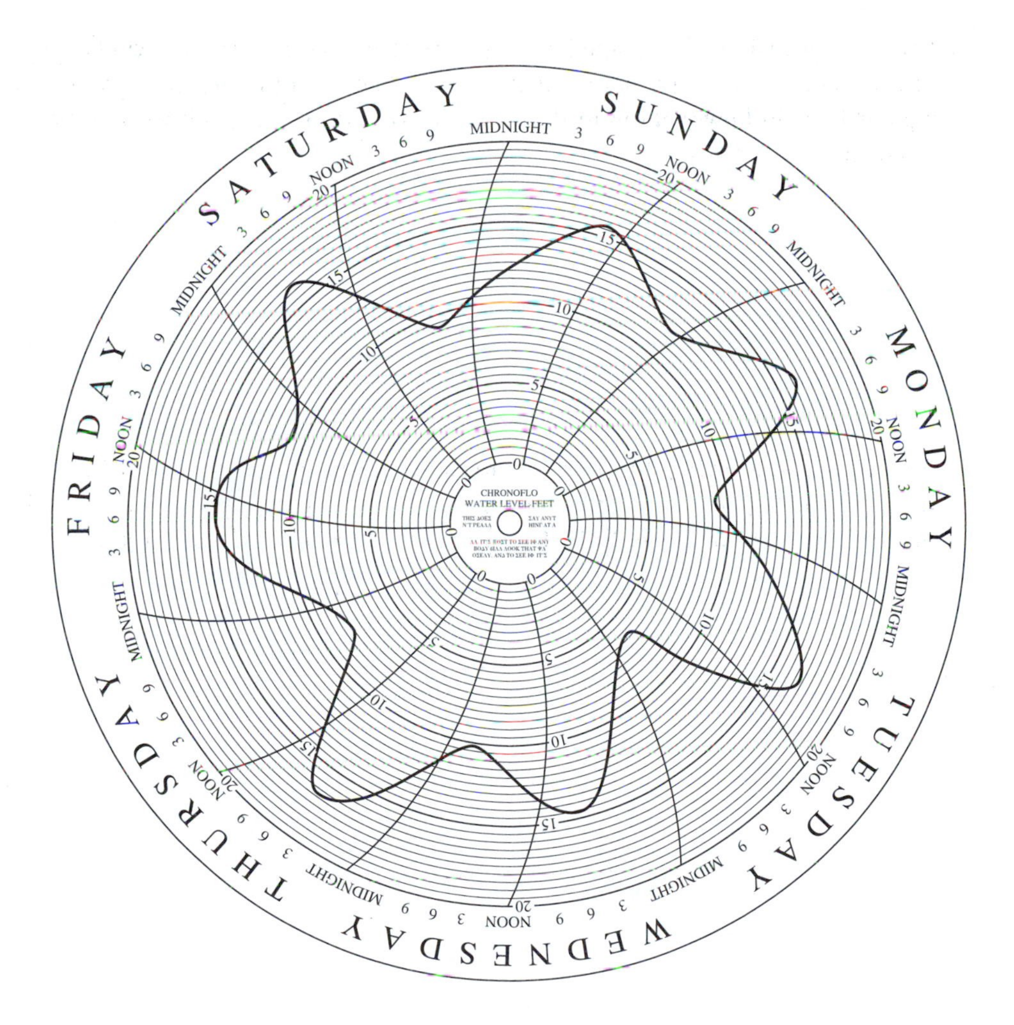

The circular graph was created when a disk of graph paper revolved slowly under a moving pen. If we were to imagine the graph "unrolled," we could see time (in hours) on the horizontal axis and water level (in feet) on the vertical axis. The resulting curve would resemble a rough sine wave. Your task is to find a function that would approximate that sine wave. You will need to take some liberties with the data. The amplitude is not constant from day to day; decide upon a representative amplitude. The high and low points don't occur at the same time each day. In order to write a sine function, you need them to be equally spaced, so feel free to assign them to 8:00 or 9:00 or whatever seems right to you. Once you've made those leaps into imprecision, you'll be able to write a sine function. (We should note that such leaps have to be taken if we're to get anywhere. If we refuse to use anything but the exact numbers, the model we get to represent our reality will be just as large and unwieldy as the reality itself.)

1. Write the formula that you came up with. Specify precisely what the independent variable represents. Explain how you decided the value for each constant.

2. Sketch a graph of your water-level function, labeling the axes and providing scales.

3. The water-level model (the idealized sine function you just wrote) also tells us something about water usage patterns in the town. Whenever more water is coming in from wells or reservoirs than people are using, the tank fills up. If people are using more water than is coming in, the tank empties.

    (a) At what time of day is the greatest amount of water being used? This is a tricky question. You need to think not about the level of water in the tank, but about the rate at which water is being drawn out of the tank.)
    (b) At what time is the least amount of water being used?

    Explain carefully how you determined your answers to the last two questions.

9.5: ALIASES—Investigating Trig Identities

In mathematics, the equals sign has three different meanings:

- It can indicate a definition, as in $f(x) = \cos(x)$. Here, the equals sign says, "The function f is defined by this mathematical expression."
- It can denote a conditional equation, as in $\cos(x) = \sin(x)$. This statement is not *always* true, but it is true for certain values of x.
- It can specify an **identity**, as in $\cos(-x) = \cos(x)$. This statement is true for *any* value of x.

Traditional trigonometry texts present pages of identities for students to memorize or to derive. Identities deserve mathematical proof before we accept them, but graphs can at least provide persuasive arguments in favor of an identity and can show that a particular relationship is *not* an identity.

Remember that, if two expressions are mathematically equivalent, they have the same graph. In this investigation, you will examine the graphs of several expressions and decide which ones are equivalent.

1. The central identity of trigonometry is $\sin^2(x) + \cos^2(x) = 1$. Explain why that statement is always true. A sketch of the unit circle would help.

2. What sort of graph should the function $y = \sin^2(x) + \cos^2(x)$ have? Make your prediction; then view the graph. You will probably have to enter it as $(\sin x)^2 + (\cos x)^2$.

3. Examine the graphs of $\sin(-x)$, $\cos(-x)$, and $\tan(-x)$ and rewrite each function in a form that uses x, rather than $-x$, as its argument.

4. Every expression on the left can be matched with an expression on the right to form an identity. Use your grapher to assist you in writing the nine identities.

$\sec^2(x) - \tan^2(x)$	$-\sin(x)$
$2\sin(x)\cos(x)$	$\sin(x)$
$\cos^2(x) - \sin^2(x)$	1
$\sin(\frac{\pi}{2} - x)$	2
$\sin(x - \frac{\pi}{2})$	$-\cos(x)$
$\cos(x + \frac{\pi}{2})$	$\cos(x)$
$\left[\cos(x) + \sin(x)^2\right] + \left[\cos(x) - \sin(x)^2\right]$	$\sin(2x)$
$\dfrac{\cos(x)}{1 - \tan(x)} + \dfrac{\sin(x)}{1 - \cot(x)}$	$\cos(2x)$
$\dfrac{\sec(x)}{\tan(x) + \cot(x)}$	$\sin(x) + \cos(x)$

9.6: DON'T *LEAN* ON ME

The Leaning Tower of Pisa is noteworthy both for its beauty and for its incline. A freestanding cylinder, it consists of eight tiers of round-arched arcades. During construction (1174–1350), an uneven settling of the ground caused the building, whose foundation was too shallow to support its weight, to lean. From the top of this famous tower, Galileo is said to have conducted experiments on gravity, the basis for the Sidney Harris cartoon in the Galileo lab.

If Galileo were to visit the Leaning Tower today, he would find a structure that leans considerably more than it did in his time. As part of ongoing efforts to preserve the fragile tower, Italian engineers monitor the angle of inclination of the structure. Measuring the tower's tilt requires some trigonometric calculation. This project asks you to use information similar to what might be supplied to the engineers to determine the angle of the tower's slant, the vertical distance from the ground to the top, and the length of its shadow.

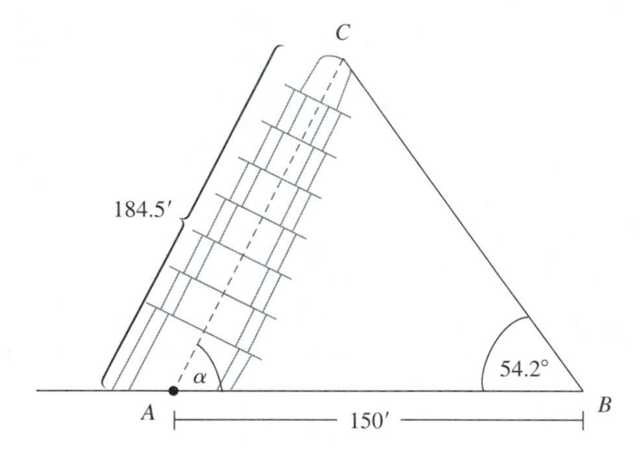

At a distance of 150 feet from the center of the tower, the angle of elevation to the top of the tower is 54.2°. The original structure would have stood approximately 184.5 feet tall if it had not started leaning prior to its completion. We will see that the situation pictured differs from a standard right-triangle problem.

1. Triangle ABC isn't a right triangle:
   (a) If $\angle A$ were a right angle, would we call this the Leaning Tower? Explain.

   (b) If $\angle C$ were a right angle, side AB would be the hypotenuse. Explain how the known side lengths make that impossible.

2. Now let's calculate α, the angle of inclination of the tower. One method of attack is to divide triangle ABC into two right triangles, as shown. We then use trigonometry to find h, and use that h-value to approximate α as the sum of α_1 and α_2.

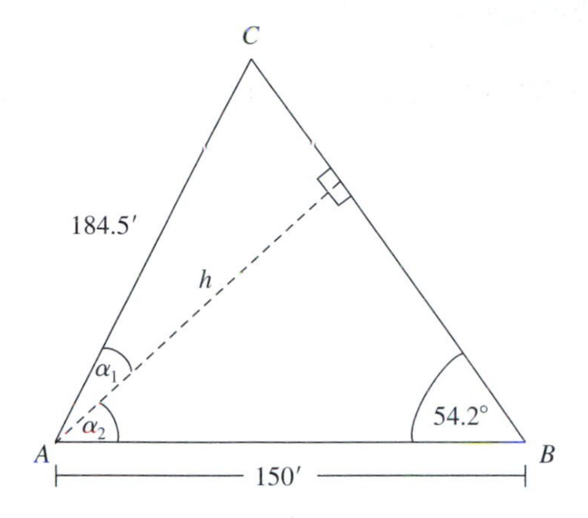

3. Next, draw the line through C that makes a right angle with the ground. The length of that line is the height of the top of the tower above the ground. What is that height?

4. If the sun were shining directly overhead, how long a shadow would the tower cast?

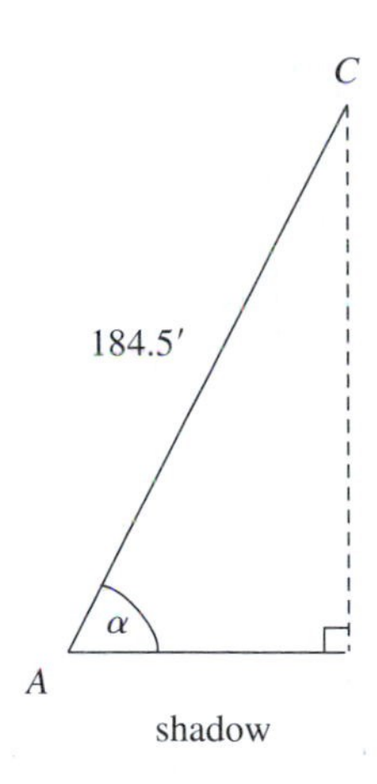

5. Pisa is actually too far north for the sun ever to shine directly overhead. Suppose, instead, that the rays of the sun come in at an angle of $60°$ with the ground. How long is the shadow cast by the tower? (*Hint:* use your results from (3) and (4).)

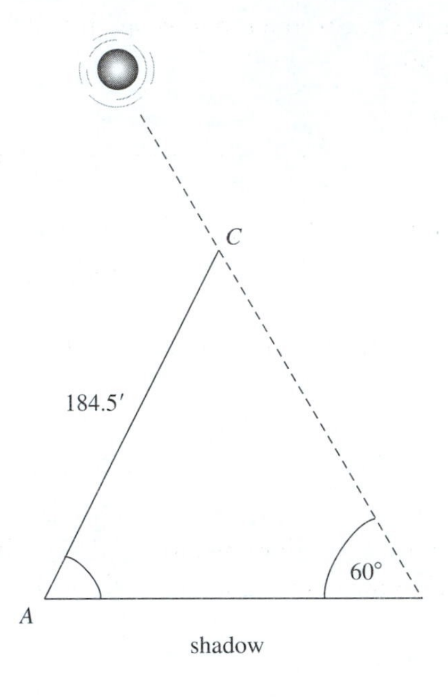

6. Find the angle of elevation of the sun that will produce a shadow twice as long as the shadow in (4).

7. Find the angle of elevation of the sun that will produce a shadow twice as long as the shadow in (5). Did you expect that angle to be half of $60°$, or something different?

9.7: HOW HIGH THE MOON?

In his treatise *On the Sizes and Distances of the Sun and the Moon,* composed around 260 B.C., Aristarchus states this observation: "When the moon is just half full, the angle between the line of sight to the sun and the line of sight to the moon is one-thirtieth of a quadrant less than a right angle." (A quadrant is 90°.)

1. What is the degree measurement of $\angle MES$, the angle made by the lines of sight to the sun and the moon, according to Aristarchus?

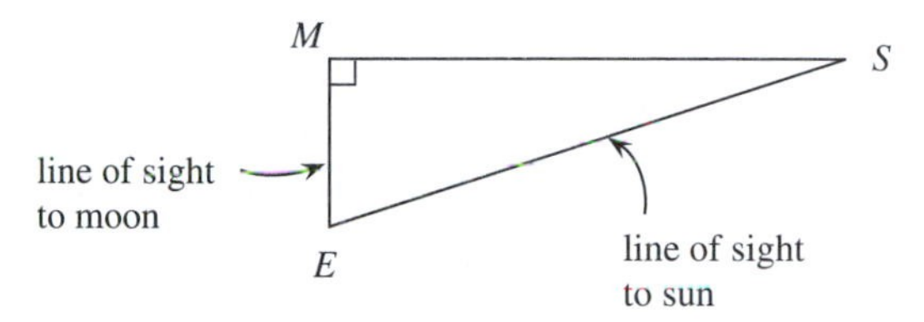

(This diagram is not drawn to scale.)

2. Use the appropriate trigonometric function to determine the ratio of $\overline{ES}$ to $\overline{EM}$. If Aristarchus had been correct, approximately how many times as far from the earth than the moon would the sun be?

3. In actuality, $\angle MES$ is approximately 89.8°. Using this present-day approximation of $\angle MES$, calculate the ratio $\dfrac{\text{distance from earth to sun}}{\text{distance from earth to moon}}$.

4. How do the modern results compare to the results using Aristarchus's observations?

Lab 10: Bordeaux

Roll Out the Barrel

PREPARATION

In the Fahrenheit lab, you examined the function $F = \frac{9}{5}C + 32$, which describes how the temperature in degrees Fahrenheit depends upon the temperature in degrees Celsius. In the Galileo lab, you saw that the function $s = 16t^2$ showed that the number of feet an object has fallen depends upon the number of seconds it has been falling.

But many quantities are determined by several inputs rather than a single one. The elevation of a point on the surface of the earth, for example, depends upon two independent variables, longitude and latitude. And the price at the pump of a gallon of gasoline in this country depends upon the wholesale price, the dealer markup, the federal tax rate, and the state tax rate. (Each of these quantities, in turn, depends upon many other factors, but that's another story.)

In the space below, write two of your own examples of functions that depend upon two or more independently varying quantities. You may or may not be able to give formulas for these functions.

The area of a square depends upon only the length of a side. We write $A(s) = s^2$, the notation $A(s)$ emphasizing the dependence of A upon a single variable s. The area of a rectangle depends upon both the length and the width. We write $A(l, w) = l \cdot w$, emphasizing the dependence of A upon the *two* variables l and w. You are already using such function notation for functions of a single variable; now we are extending it to functions of several variables.

Here are some diagrams that appear inside the front cover of a typical precalculus text:

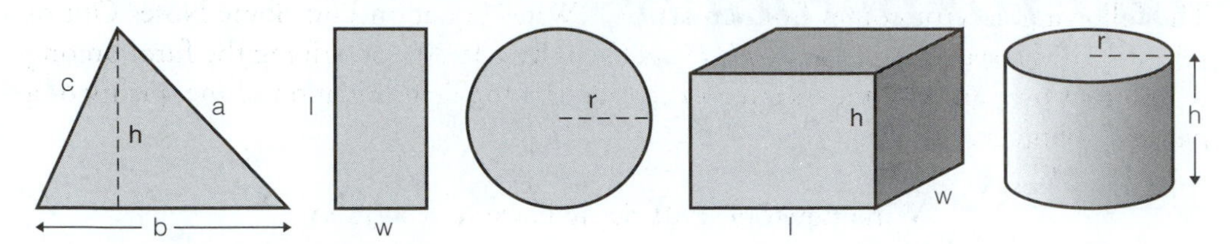

Write the area functions for the triangle, the rectangle, and the circle, expressing them in function notation. How many independent variables appear in each function?

Write the perimeter (circumference) functions for the same shapes. How many variables does each function have?

For the two solids, write the surface area and volume formulas, expressing them in function notation. How many independent variables occur in each?

THE BORDEAUX LAB

The following excerpt comes from an article, "Wine Equation Puts Some Noses Out of Joint," by P. Passell, in the *New York Times* (March 4, 1990), describing the furor among wine critics over an economist's attempt to quantify the determination of the quality of a particular vintage.

Wine Equation Puts Some Noses out of Joint

> Calculate the winter rain and the harvest rain (in millimeters). Add summer heat in the vineyard (in degrees centigrade). Subtract 12.145. And what do you have? A very, very passionate argument over wine.
>
> Professor Orley Ashenfelter, a Princeton economist, has devised a mathematical formula for predicting the quality of red wine vintages in France. And the guardians of tradition are fuming.
>
> ... It is widely agreed that weather influences wine quality. What few understand, Ashenfelter argues, is that a mere handful of facts about the local weather tell almost all there is to know about a vintage.[1]

The Bordeaux Equation that Professor Ashenfelter uses to predict wine quality is

$$Q = 0.0117\,WR - 0.00386\,HR + 0.6164\,TMP - 12.145$$

where

WR = winter rain (October through March) in millimeters

HR = harvest rain (August through September) in millimeters

TMP = average temperature during growing season (April through September) in degrees Celsius

Q = a number Ashenfelter calls the quality index (the higher the index, the better the wine)

Rewrite the Bordeaux Equation in function notation. How many independent variables are there? What are they?

You have learned to distinguish between linear and nonlinear functions when a single variable is involved. The same concept can be extended to functions of more than one variable. A function having the form

$$f(x_1, x_2, \ldots, x_n) = a_0 + a_1x_1 + a_2x_2 + \cdots + a_nx_n$$

where the a_i are constants and the x_i are the independent variables, is called a linear function of n variables. (Notice that no variable is multiplying or dividing another and that each variable is raised to the first power only.)

Refer to the functions you wrote for the preparation section. Compare with those of your partners to be sure you are in agreement. Decide which of the functions are linear.

Examine the functions from the Fahrenheit and the Galileo labs:

$$F = \frac{9}{5}C + 32 \quad \text{and} \quad s = 16t^2$$

Is either one linear? What does its graph look like? How can you tell from the *formula* that one isn't linear? How can you tell from its *graph*?

Now, back to the Bordeaux Equation: in order to determine Q for a given year, how many inputs are required? Is Q a *linear* function of those variables or not?

Which do you think would contribute more to the quality of a particular vintage: an increase of one degree Celsius in the average temperature during the growing season, or an increase of 10 millimeters in the amount of winter rain? Which would you say is more desirable to the wine grower, winter rain or harvest rain? How did you determine your answers to the last two questions?

In 1988, the winter rains were about average, the temperature during the growing season was above average, and August and September were unusually dry. Since the flavor of the wine does not develop fully in the bottle for at least ten years, a definitive testing cannot take place before 1998. If Orley Ashenfelter had the opportunity in 1997 to buy the 1988 Bordeaux cheaply, do you think he would do so? Justify your answer.

Suppose that the winter rain and average temperature in another year are about the same as those in 1989, but September of that year is a very rainy month. Does Q increase or decrease with the additional September rainfall? How do you know?

Let's consider the difficulties inherent in drawing graphs for a function of more than one input variable. We are accustomed to plotting the independent variable on the horizontal axis and the dependent variable on the vertical axis. What shall we do if we have more than one independent variable?

Topographical maps are one solution. They plot altitude as a function of both latitude and longitude by means of **level curves** for different elevations, while the input variables monopolize the horizontal and vertical axes.

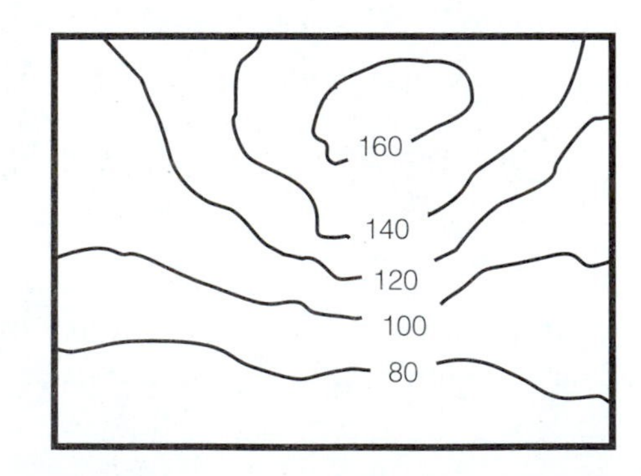

The same information could have been presented using different shades of color to indicate different elevations. We see examples of this kind of graphical presentation in an atlas. Even so, these solutions take care of only *two* independent variables. With three or more variables, we need to be either extremely creative (and artistically gifted), or able to take an easier way out.

Perhaps the most straightforward method is to pretend that all of the variables except one are constant, and then study what happens to the function when we change just that one variable. In other words, we convert the multivariable problem to a single-variable problem. That's the approach we'll use in this lab.

Suppose that *WR* is 583 millimeters. Use your grapher to examine how *Q* varies with *HR* for three different average temperatures: 15, 18, and 20 degrees Celsius.

Sketch these three graphs on the same set of axes. Label the axes and identify each graph. Be sure that your sketch includes the entire portion of the graph that makes sense in context. (What rainfall amounts might be reasonable for a two-month period? Consider the units in which *HR* is measured; how large is a millimeter?)

Keep in mind that the purpose of a graph is to convey information. Does your graph give a visual image of how *Q* depends upon *HR* for various temperatures? Should higher values of *TMP* be considered as well? Recall what temperatures *TMP* represents and discuss whether you ought to include any more graphs in your sketch.

Then, suppose *TMP* = 16 degrees. Use your grapher to examine how *Q* varies with *WR* for three different amounts of harvest rain: 50, 100, and 170 mm.

Sketch these two graphs on another set of axes, again making sure that you show a sensible portion of the graph. Would a person examining your graphs understand what makes the quality of wine go up and what makes it go down?

Recall that the range of a function is the set of output values. Examine your graphs. Do you think the range of this mathematical model could include negative numbers? Explain.

In precalculus, you study various types of scales. The Fahrenheit lab presented two different scales for measuring temperature. In the Earthquakes lab, you learned the value of using logarithms to rescale seismographic data that varied widely in magnitude.

In this lab, we are using a scale to predict the quality of wine. You might have noticed that the numbers generated by the Bordeaux Equation for realistic temperatures and amounts of rainfall are all "in the same ballpark." In other words, most of the *Q*-values likely to be produced by conditions in a temperate climate lie in a cluster around 3 and 4. That clustering of values makes it difficult to distinguish the quality of one wine from that of another. A *Q*-value of 4.4, for example, doesn't sound much better than a *Q*-value of 4.2. To spread out the quality measurements, Ashenfelter treats them as *exponents*—that is, as *logarithms*—whose base is the irrational number *e*. Obtain decimal approximations for the numbers $e^{4.4}$ and $e^{4.2}$, and you'll have a sense of how much better a 4.4 wine is than a 4.2 wine.

The constant −12.145 in the Bordeaux Equation was chosen so that, when the variables representing winter rain, average temperature, and harvest rain were replaced by their 1961 values, *Q* turned out to be 4.6052. Why 4.6052? Whip out your calculator again and ask it to approximate $e^{4.6052}$. What's the result? The 1961 vintage was spectacular, and it has become the norm against which all other Bordeaux wines are measured. Do you see why the author of the Bordeaux Equation would want an outcome of 4.6052 for that year?

The ratings in the following table are predictions only, because Bordeaux wine must mature for at least ten years for its quality to develop fully, and the evaluations were made sooner than that. The table shows a comparison between the results of Orley Ashenfelter's calculations for two types of Bordeaux wines and the ratings of two prominent wine critics.

Year	Ashenfelter (scale of 0 to 100)	Hugh Johnson (scale of 0 to 10)	Robert M. Parker, Jr.
1987	38	3 to 6	Pleasant, soft, clean, fruity
1986	23	6 to 9	Very good, sometimes exceptional
1985	65	6 to 8	Soft, fragrant, very good
1984	33	4 to 7	Austere, mediocre quality
1983	76	6 to 9	Superior to 1981, rarely achieves greatness
1982	56	8 to 10	Most complex and interesting wines since 1961
1981	42	5 to 8	Lacks generosity and richness
1980	28	4 to 7	Light and disappointing

Ashenfelter's calculations, based on his model, suggest that 1989 Bordeaux wines will prove to be the greatest of the century. Would you be willing to invest in wine futures for the 1989 Bordeaux vintages? That is, do you think the wine equation is a good model for determining the quality of a particular vintage? Support your answer with references to the table.

THE LAB REPORT

Functions requiring two or more inputs are very common in fields ranging from weather forecasting to economics. (These functions are called **multivariable functions**.) Give three examples of functions used as models that require two or more independent variables. State what the variables measure. Include at least one linear function.

Explain how you have learned to distinguish linear functions from nonlinear ones when there is more than one input variable. Discuss the difficulties of graphing a multivariable function and describe the method you used to produce graphs for the Bordeaux Equation. Include, as illustrations, the graphs you drew.

Explain how you were able to determine whether each variable had a favorable or unfavorable effect on the quality of the wine, and explain how you could tell whether one variable had a stronger effect than another. Give your group's evaluation of the Bordeaux Equation as a model for predicting wine quality.

Finally, describe the effects you observed when you treated the Q-values as logarithms.

Section 10—Projects and Explorations

10.1: QUILTS

This project is a follow-up to Lab 10

Here you see three traditional American quilt patterns:

Mohawk Trail

Friendship Star

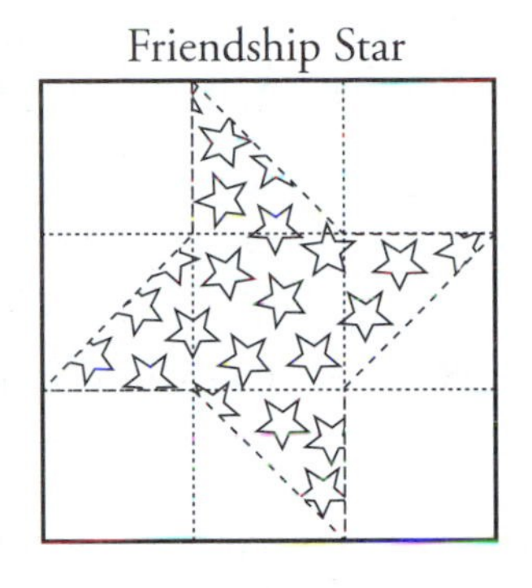

Drunkard's Path

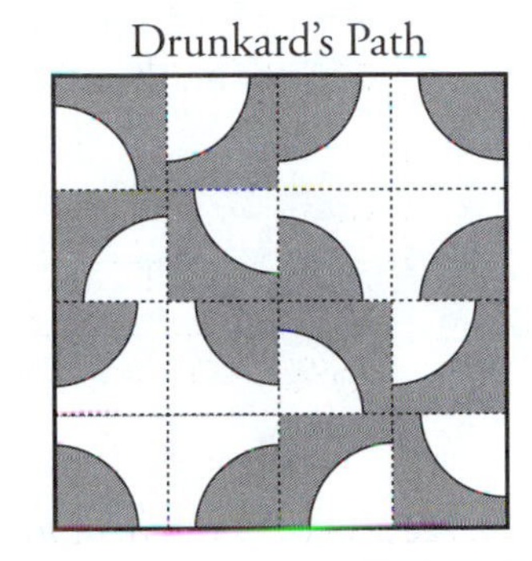

1. Sophie Scott, an avid quiltmaker, decides to incorporate 10-inch squares from all three patterns in a quilt, together with some plain square patches. These last are obviously the easiest to make; in fact she can make eight plain patches in an hour. By contrast, she can make only two Friendship Star squares in an hour. In *hours per square,* what is Sophie's rate for the plain squares? What is her rate for the Friendship Star squares?

2. In the Drunkard's Path pattern, a curved piece is cut out of a small dark $2\frac{1}{2}$-inch square and replaced with an identical light one. This pattern requires much more work because the curves are tricky to sew; it takes Sophie three hours to make a 10-inch Drunkard's Path square. The Mohawk Trail pattern is even more difficult, because each curved portion is composed of three separate pieces. She needs five hours to make each such 10-inch square. What is her rate, in hours per square, for the Drunkard's Path squares? For the Mohawk Trail squares?

The Quilting Function

The length of time that it takes Sophie to complete a quilt depends upon the number of squares of each type.

Let s be the number of Friendship Star squares in the quilt.
Let d be the number of Drunkard's Path squares in the quilt.
Let m be the number of Mohawk Trail squares in the quilt.
Let p be the number of plain squares in the quilt.

3. Using the rates you computed in (1) and (2), write a function of these four variables, $t(s, d, m, p)$, that gives the time required to construct all the squares for the quilt in terms of the numbers of squares of the different types. Is your function linear? Explain, using the algebraic formula for the function to help you decide.

4. You have two sets of values to keep track of: the domain variables s, d, m, p, which count the number of squares of each type; and the four rates, which tell how long it takes to make a square of a particular type. Let's focus now on the domain variables, and suppose that Sophie plans to make a 60-inch by 80-inch quilt for a queen-size bed. What is the total number of squares she needs?

5. Use that number and all four domain variables to write an equation.

6. The equation you just wrote imposes a restriction on the sum of the domain variables. You could solve that equation for any one of the four variables. For now, solve the equation for d to obtain an expression for d in terms of the other three variables.

7. Go back to the *time* function you wrote and substitute the expression for d. You should now have t as a function of just the three variables p, s, and m.

8. Is the new function linear? Explain your answer in terms of the formula for the function.

9. The Mohawk Trail squares are so difficult that Sophie decides to use only four of them, placed at the very center of the quilt. Use your grapher to see how the dependent variable t varies with p for s-values of 10, 20, and 30.

10. Sketch these three graphs on the same set of axes, labeling the axes, providing scales, and identifying each graph. Use your graphs to complete the assignment.

11. Discuss similarities and differences between the graphs.

12. As p increases, what happens to t? Explain why this makes sense in context.

13. For a given value of p, which of your three graphs has the smallest t-value? (This means you need to fix the value of p and compare the three graphs with one another at that p-value. You could do this visually by sketching a vertical line that crosses all three of your graphs.) Explain why the answer makes sense in the context of the quilting problem. (This requires a little hard thinking!)

14. Recall that the range of a function is the set of output values. Estimate the range of t if we restrict Sophie to 10-inch squares of the four patterns and a 60-inch by 80-inch quilt, but impose no other restrictions.

10.2: DIMENSIA

In making containers of various sizes and shapes, manufacturers should understand the relationship between the dimensions they choose for the container and the amount of material needed. Consider, for example, an open cardboard box with a square bottom. In this project, you will analyze how the amount of cardboard used is affected by changing one dimension or the other.

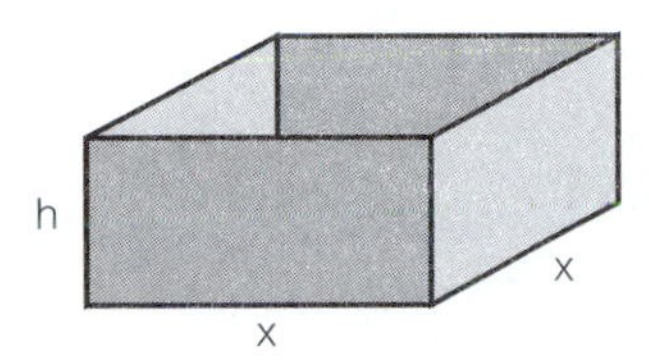

1. Let $S(x, h)$ be the surface area of the box. Write a formula for $S(x, h)$, the relationship between the amount of cardboard material in the box and the dimensions of the box. (Remember that the box has a bottom but no top.)

Holding the Base Constant

2. Suppose the base of the box measures 10 inches by 10 inches; that is, $x = 10$. Rewrite the surface-area function and graph $S(10, h)$, the relationship between the amount of cardboard and the height of the box. What kind of function is $S(10, h)$?

3. How much additional cardboard will be required if the height increases from 2 inches to 3 inches? From 6 inches to 7 inches? From n inches to $n + 1$ inches? Tell how you decided.

4. Now change the size of the base, letting x be 11 inches. Rewrite the formula for S and graph $S(11, h)$.

5. How much additional cardboard will be required with this larger box if the height increases from 2 inches to 3 inches? From 6 inches to 7 inches? From n inches to $n + 1$ inches?

6. Change the size of the base again, letting x be 12 inches. Write the formula for $S(12, h)$ and answer the questions in (5) for this box.

7. Sketch the graphs of $S(10, h)$, $S(11, h)$, and $S(12, h)$. Label axes, show scales, and identify each graph.

8. The function $S(x, h)$ has a different rate of change for each fixed value of x.
   (a) For $x = 10$, the rate of increase in surface area is ______ square inches of area per inch of height.
   (b) For $x = 11$, the rate of increase in surface area is ______ square inches of area per inch of height.
   (c) For $x = 12$, the rate of increase in surface area is ______ square inches of area per inch of height.

   As the value of x increases, describe what happens to the *rate of increase* in surface area with respect to height.

Holding the Height Constant

9. Suppose the box has a height of 1 inch. Write the formula for $S(x, 1)$, the relationship between the amount of cardboard and the length of a side of the base. View its graph. What type of function is $S(x, 1)$?

10. How much additional cardboard would be required to make a box whose base is an 11-inch square instead of a 10-inch square? A 12-inch square instead of an 11-inch square? A 13-inch square instead of an 12-inch square?

11. Suppose you were to change the length of the base from 12.7 inches to 13.7 inches. Without doing any calculations, state whether the additional amount of cardboard would be the same as, more than, or less than the additional amount needed when you changed that dimension from 12 inches to 13 inches. How did you decide?

12. Write the algebraic formulas for $S(x, 1)$, $S(x, 2)$, and $S(x, 2)$. Sketch their graphs, label the axes, show scales, and identify each graph.

13. Which of the three graphs appears to be rising most steeply? Calculate the amount of additional cardboard needed when the base dimension changes from 10 inches to 11 inches for boxes of height 1 inch, 2 inches, and 3 inches. In other words, compute three separate *average rates of change.*

14. Compare your two sets of graphs. What can you learn from them about boxmaking?

10.3: SKINTIGHT

How much skin covers your body? One mathematical model for estimating your body's surface area is this:

$$A(w, h) = 15.63\, w^{0.425}\, h^{0.725}$$

where surface area A is measured in square inches, weight w is in pounds, and height h is in inches.

1. Find $A(125, 64)$, and write a sentence saying what it means.

2. Predict the body surface area of a person 5′8″ tall who weighs 185 pounds.

3. If we assume that the model works well in predicting body surface area for adults, decide upon a reasonable domain and range for the model. (Remember to consider both domain variables.)

4. Determine whether or not A is a linear function of w and h, and explain how you can tell this from the formula for A.

5. Draw a graph of the relationship between body surface area and height for individuals who weigh 130 pounds, keeping h within the limits of the domain you specified in (3). Put scales on the axes and label the graph $w = 130$.

6. For 130-pound people, does body surface area increase or decrease as height increases? How do you know this from the graph?

7. On the same axes, graph the relationship between body surface area and height for people weighing 175 pounds. Label your graph $w = 175$.

8. For a *fixed* height, will body surface area increase or decrease as w increases from 130 pounds to 175 pounds? Explain how the graphs provide this information.

Lab 11: Bézier Curves

Draw your own conclusions

PREPARATION

A television commercial opens with the image of a car on a computer screen. As you watch, the car's design is modified to give passengers more leg room, and then further modified to make the care aerodynamically more efficient. The end result of all this tinkering is, of course, the ideal car, the one that the manufacturer hopes you'll rush out and buy.

With computer-aided design (CAD) software, car designers can work interactively with the computer to visualize and then refine their designs for the new models. Because each car design consists of many different curves pieced together, CAD software must enable designers to produce a wide variety of curves on a computer screen.

Imagine that you are part of a team in charge of designing next year's model. Sketch a side view of the car you envision.

In another color, shade at least three different curve segments in your design that resemble graphs you have studied this semester. (One of your curves might be a straight line. In mathematics, a line is considered to be one type of curve.) Identify the family of functions or relations to which each of your selected curves belongs.

A computer needs precise mathematical descriptions for all the curves in a design before it can display that design on the screen. In the 1960s, P. Bézier and P. de Casteljau, engineers working in the French automotive industry, independently developed a mathematical method for designing a wide variety of curves. Initially, the results were considered manufacturing secrets; today, **Bézier curves** (the curves produced by this method) are widely used in CAD. (Even though the work of de Casteljau preceded that of Bézier, the latter was published much more widely and so Bézier gets the credit.) The mathematical descriptions of Bézier curves rely on parametric equations, which allow us to describe curve segments that could not be represented by a function, such as a sideways parabola or a circular loop.

Although you've probably never been part of an automotive design team, you might have used a computer drawing program such as Microsoft *Paintbrush*. With the *Paintbrush* curve tool, you can create a variety of curves by clicking on two, three, or four **control points** that you choose. Have you ever wondered how the program uses your control points in determining precisely which curve it will draw? (No, it doesn't just connect all the points.)

If you specify two control points, *Paintbrush* draws the line segment that connects the two points, a first-degree Bézier curve. For example, suppose that you choose the control points $P_0(3, 2)$ and $P_1(5, 7)$. Plot these points in a Cartesian (x, y) plane and then sketch the line segment connecting P_0 to P_1. Any computer drawing program needs a mathematical description in order to draw your first-degree Bézier curve. Several different mathematical formulas can describe this line segment.

- Describe the line segment in function notation. Be sure to restrict the domain of the function so that its graph is precisely the line segment that you've sketched and not the entire line.

- Describe the line segment parametrically: determine a set of parametric equations that describe uniform (constant rate) motion along the line segment, starting at $P_0(3, 2)$ when $t = 0$ and ending at $P_1(5, 7)$ when $t = 1$.

You now have two mathematical descriptions for the first-degree Bézier curve determined by control points P_0 and P_1. As you complete the Bézier lab, you will learn the answers to related questions:

- How will the shape of the Bézier curve change if one or two additional control points are chosen?
- Mathematically, how can you describe Bézier curves determined by two, three, or four control points?

THE BÉZIER LAB

In this lab, you'll learn the mathematical rules used by computer drawing software such as *Paintbrush* and more sophisticated CAD programs to create Bézier curves of the first, second, and third degree. In addition, you'll use your grapher to investigate the various shapes that these curves can produce.

First-Degree Bézier Curves

A first-degree Bézier curve is completely determined by *two* control points. If the control points are $P_0(x_0, y_0)$ and $P_1(x_1, y_1)$, then the first-degree Bézier curve is simply the line segment joining P_0 and P_1. In the preparation, you wrote two mathematical descriptions for the Bézier curve determined by $P_0(3, 2)$ and $P_1(5, 7)$. Now we need a mathematical description using parametric equations for the line segment connecting any two arbitrary control points $P_0(x_0, y_0)$ and $P_1(x_1, y_1)$.

As before, we look for a pair of parametric equations to describe motion at a constant rate along the line segment joining $P_0(x_0, y_0)$ and $P_1(x_1, y_1)$, starting at P_0 when $t = 0$ and ending at P_1 when $t = 1$. Recall that parametric equations of the form

$$x(t) = a + bt, \qquad y(t) = c + dt$$

describe motion at a constant rate along a line. Determine the values of the coefficients a and b by substituting $t = 0$ and $t = 1$ into $x(t)$

$$x(0) = a + b(0) = x_0$$
$$x(1) = a + b(1) = x_1$$

and then solving for a and b in terms of x_0 and x_1. Next, find c and d in terms of y_0 and y_1. Write your mathematical description for the first-degree Bézier curve determined by $P_0(x_0, y_0)$ and $P_1(x_1, y_1)$:

$$x(t) = \_\_\_\_ + \_\_\_\_\_\_\_\_ t \qquad 0 \le t \le 1 \qquad (1)$$
$$y(t) = \_\_\_\_ + \_\_\_\_\_\_\_\_ t \qquad 0 \le t \le 1$$

Frequently, the mathematical description for the first-degree Bézier curves determined by the control points $P_0(x_0, y_0)$ and $P_1(x_1, y_1)$ is given in this form:

$$x(t) = (1 - t)\,x_0 + tx_1 \qquad 0 \le t \le 1 \qquad (2)$$
$$y(t) = (1 - t)\,y_0 + ty_1 \qquad 0 \le t \le 1$$

As you will see later in this lab, this pattern generalizes more easily to higher-degree curves than the pattern in (1). Do the algebra that allows you to express your equations from (1) in the form shown in (2).

Notice that the formulas for the x- and y-coordinates in (2) have the same form: $(1-t)$ times the coordinate from P_0 plus t times the coordinate from P_1. If we let $B(t)$ represent the point on the line segment at time t, we can summarize the two equations in (2) with a single equation:

$$B(t) = (1-t)\,P_0 + t\,P_1 \qquad 0 \le t \le 1 \tag{3}$$

This means that $B(t) = (x(t), y(t))$, where

$$x(t) = (1-t)\,x_0 + tx_1 \quad \text{and} \quad y(t) = (1-t)\,y_0 + ty_1$$

Viewed in this form, the first-degree Bézier curve is the result of using the x- and y-coordinates of both control points with two first-degree polynomials in t: $(1-t)$ and t. Later, we will see how this pattern extends to three or four control points.

Here are two computer-generated first-degree Bézier curves. Coordinate axes have been added so that you can identify the control points.

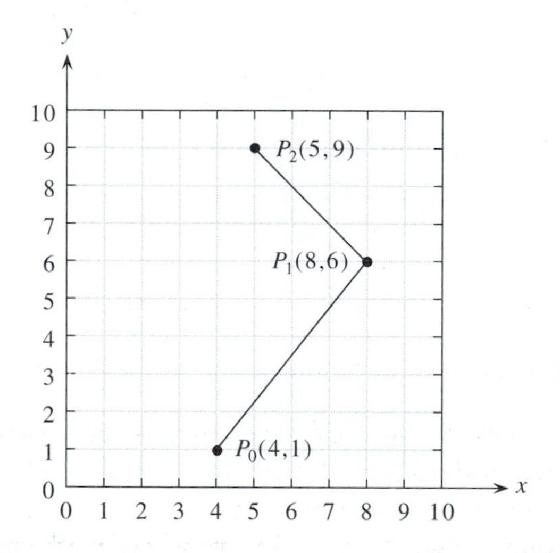

Following the form given in (2), write the mathematical description for $B_0(t)$, the first-degree Bézier curve connecting P_0 to P_1, and $B_1(t)$, the first-degree Bézier curve connecting P_1 to P_2. Then use your grapher to verify that these equations produce the same picture.

Second-Degree Bézier Curves

You just used three control points, but you produced two separate first-degree curves. Now we will generate a single curve controlled jointly by all three points. For comparison, let's use the same points: $P_0(4, 1)$, $P_1(8, 6)$, and $P_2(5, 9)$. We'll let $B(t)$ represent the resulting curve:

$$B(t) = (x(t), y(t)) = (1-t)\,B_0(t) + t\,B_1(t) \tag{4}$$

The difference between form (4) and form (3) is that $B_0(t)$ and $B_1(t)$ are themselves parametric equations rather than simply coordinates of points. In particular,

$$x(t) = (1 - t)(x\text{-coordinate from } B_0) + t(x\text{-coordinate from } B_1)$$

Remember that each of those x-coordinates is an entire algebraic expression

$$\begin{aligned} x(t) &= (1 - t)[(1 - t)(4) + t(8)] + t[(1 - t)(8) + t(5)] \\ &= (1 - t)^2(4) + 2t(1 - t)(8) + t^2(5) \\ &= (1 - t)^2(x\text{-coordinate of the first control point}) \\ &\quad + 2t(1 - t)(x\text{-coordinate of the second control point}) \\ &\quad + t^2(x\text{-coordinate of the third control point}) \end{aligned}$$

and

$$y(t) =$$

In the space provided, write out the equation for $y(t)$.

Notice that the formulas for $x(t)$ and $y(t)$ use the x- and y-coordinates of the three control points with three quadratic polynomials in t: $(1 - t)^2$, $2t(1 - t)$, and t^2. The result $B(t)$ is an example of a second-degree Bézier curve.

With the aid of your grapher, view the curve whose mathematical description you completed in (4). The correct curve should begin at $P_0(4, 1)$, bend in the direction of $P_1(8, 6)$, and end at $P_2(5, 9)$.

Every second-degree Bézier curve is determined by three control points, $P_0(x_0, y_0)$, $P_1(x_1, y_1)$, and $P_2(x_2, y_2)$. If, as we did before, we let $B_0(t)$ and $B_1(t)$ be the first-degree Bézier curves connecting P_0 to P_1 and P_1 to P_2, respectively, then we obtain the second-degree Bézier curve by combining $B_0(t)$ and $B_1(t)$ as follows:

$$B(t) = (1 - t)\, B_0(t) + t\, B_1(t) \qquad 0 \le t \le 1 \tag{5}$$

Substitute the correct parametric expressions for $B_0(t)$ and $B_1(t)$, and do the algebra necessary to verify that we can express the combination of parametric equations in (5) as

$$B(t) = (1 - t)^2 P_0 + 2t(1 - t)P_1 + t^2 P_2(t) \tag{6}$$

The second-degree Bézier curve is thus the result of combining the three control points P_0, P_1, and P_2 with the same three quadratic polynomials in t: $(1 - t)^2$, $2t(1 - t)$, and t^2.

Return to the three points you were using earlier: $P_0(4, 1)$, $P_1(8, 6)$, and $P_2(5, 9)$. Let's find out what happens when we change the middle control point. For each change in P_1, write a new mathematical description of the second-degree Bézier curve determined by the original P_0 and P_2 with the new P_1. Then use your grapher to examine the shape of the curve.

- Move P_1 vertically by replacing $(8, 6)$ with, say, $(8, 11)$ or $(8, 0)$.
- Move P_1 to the left or to the right of $(8, 6)$.
- Move P_1 diagonally from $(8, 6)$.

Describe the shapes you observed. Can a second-degree Bézier curve ever be a line? If so, what must be true of the control points? If not, why not?

Third-Degree Bézier Curves

As you might guess, third-degree Bézier curves are determined by four control points, P_0, P_1, P_2, and P_3. The mathematical description of this curve

$$B(t) = (1-t)^3 P_0 + 3t(1-t)^2 P_1 + 3t^2(1-t)P_2 + t^3 P_3 \qquad 0 \le t \le 1 \qquad (7)$$

is the result of combining four control points and using four cubic polynomials in t. This formula comes from algebraic calculations similar to the ones you used to find $B(t)$ for the second-degree curve, but we won't make you struggle through this one.

Using (7), find a set of parametric equations to describe the Bézier curve determined by the control points $P_0(2, 1)$, $P_1(6, 5)$, $P_2(4, 0)$, and $P_3(10, 3)$. With the aid of your grapher, sketch the curve. How many of the control points lie *on* the curve?

Next, investigate the effect of changing the two middle control points, P_1 and P_2. For each change, write the mathematical description of the new third-degree Bézier curve. Then examine the shape of the curve with your grapher. (Notice that, for the four given points, the two middle ones lie on opposite sides of the line segment joining the first and last points.)

- Move P_1 higher and P_2 lower than their current positions.
- Move P_2 so that it lies above the direct path from P_0 to P_1.

Continue experimenting until you have a good idea of the variety of shapes possible for a third-degree Bézier curve that starts at $(2, 1)$ and ends at $(10, 3)$.

Now it's time to test what you have learned. Look at the two sets of control points pictured. Without the aid of a grapher, make a rough sketch of the third-degree Bézier curves determined by these points.

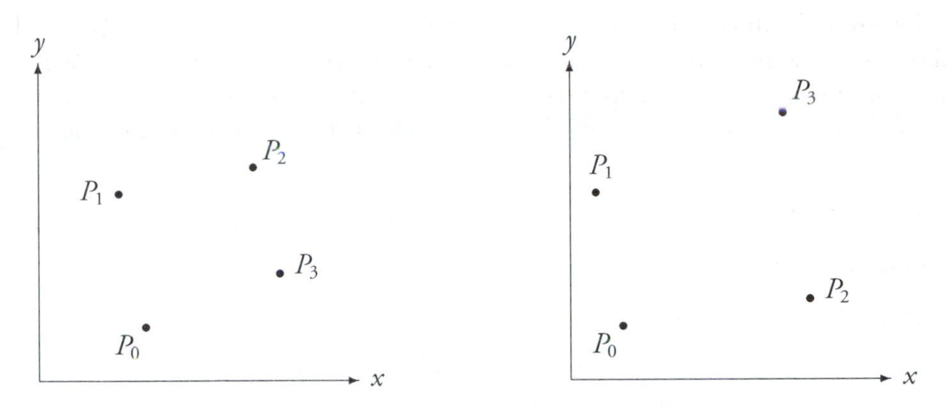

What is the minimum number of control points needed to produce each of the following Bézier curves? Mark the approximate location of each control point.

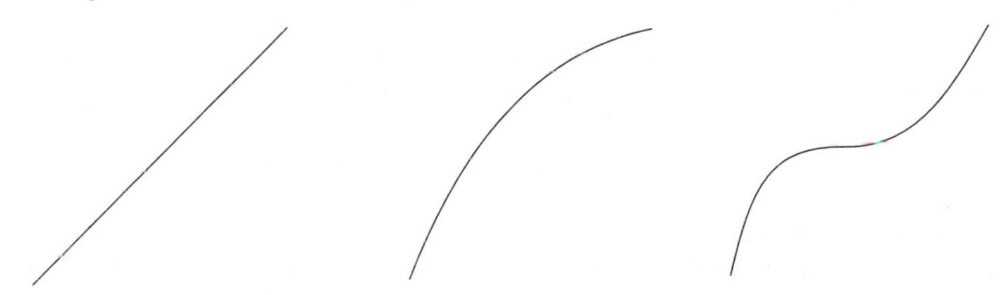

Let's summarize what you've learned about Bézier curves. First-degree, second-degree, and third-degree Bézier curves are determined by two, three, and four control points, respectively, and are defined by parametric equations that use polynomials. A first-degree Bézier curve consists of the line segment connecting the two points. Each additional control point makes it possible for the curve to bend in another direction. You've been able to create a variety of shapes using only first-, second-, and third-degree Bézier curves. Some sophisticated CAD software allows for Bézier curves of even higher degrees. A click of the mouse adds or moves a control point, allowing designers to experiment with and refine a profile until they are satisfied.

THE LAB REPORT

Discuss what you have learned about Bézier curves.

- Explain how the mathematical descriptions of first- and second-degree Bézier curves arise. Show the missing algebra in the construction of the formulas.
- Provide sketches of several first-, second-, and third-degree Bézier curves. Give mathematical descriptions for the curves you have sketched.
- With a series of sketches, show how a change in one or more of the middle control points of a second- or third-degree Bézier curve alters the shape of the curve.

Conclude your report with a summary of the shapes of curves you could draw using a computer drawing program capable of producing first-, second-, and third-degree Bézier curves. If you have access to such software, feel free to experiment and to include in your report samples of curves you have created.

Section 11—Projects and Explorations

11.1: WHAT GOES AROUND COMES AROUND

Use square scaling for this project, so that circles look like circles and not like ellipses. If you have a choice between *degree mode* and *radian mode*, use radian mode. For all graphing in (1) through (7), set the time interval as $0 \leq t \leq 2\pi$ (or, approximately, $0 \leq t \leq 6.3$). Assume that time is measured in seconds. If your grapher calls for a time increment, use 0.1 second.

1. Suppose that the motion of an object is governed by the parametric equations

$$x = \cos(t), \quad y = \sin(t)$$

   Draw a graph of the path of motion. Label the position of the object every 0.5 second. From the spacing of the labeled positions, does it appear that the object is traveling at a constant rate? Explain.

2. The central identity of trigonometry is $\cos^2(t) + \sin^2(t) = 1$. Use this identity to find a single equation, not involving t, to describe the path of the object in (1). What geometric shape does this equation represent?

3. Now alter the basic equations from (1) in three ways.
   (a) Add a constant c to the x-equation. Use several values for c, both positive and negative. Experiment until you can describe how the addition of a constant to the x-equation affects the object's path.

   (b) Use the original x-equation, but add a constant k to the y-equation. Again, experiment; how does k affect the path?

   (c) Add constants to both equations at the same time. How is the path of the object affected?

   (d) Does the addition of one or more constants change the rate of travel? Explain.

4. Using this set of parametric equations,

$$x = \cos(t) + 2, \quad y = \sin(t) + 1$$

write a single equation for the path that does not involve the variable t. What geometric shape does this equation represent?

5. This time, alter the equations in (1) by multiplying the x-equation by a positive constant k. How does this modification alter the path of motion? In your investigation, choose at least two values for k that are greater than 1 and at least two that are between 0 and 1. Draw sketches illustrating what you learned.

6. Use the fundamental trigonometric identity in (2) to write a single equation, not involving t, for the path of an object moving according to the parametric equations

$$x = 3\cos(t), \quad y = \sin(t)$$

What geometric shape does this equation represent?

7. What happens to the path of the object if both the x- and y-equations in (1) are multiplied by the same positive constant k? Sketch graphs of the paths of motion for several choices of k.

8. Instead of multiplying by a constant, suppose we multiply both equations in (1) by the parameter t, as shown:

$$x = t\cos(t), \quad y = t\sin(t)$$

Using the time interval $0 \leq t \leq 20$, plot the position of the object at every second. (Remain in square scaling.) Sketch the curve you see.

9. The graph you have sketched is called the **spiral of Archimedes**. As the value of t increases, does the rate at which the object is moving along the spiral increase, decrease, or stay the same? Explain how you can tell.

11.2: PARAMETRIC IMPERSONATIONS

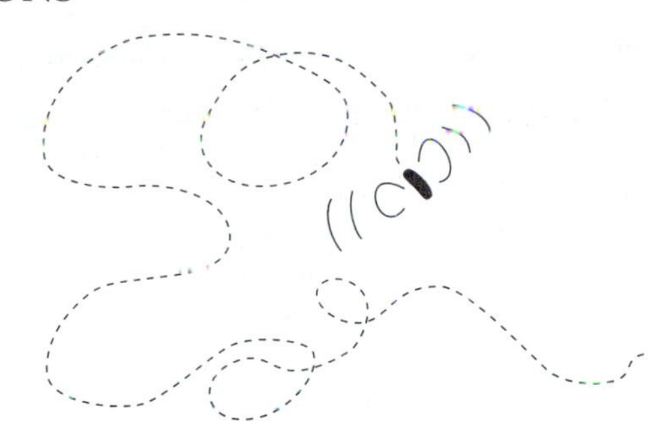

The path of a fly's flight can be filled with loops and strange twists. The equations here might recreate some of the more memorable flights that you've witnessed. Graph each set of equations. Determine whether or not the path eventually folds back upon itself and starts to repeat the pattern. If it does, approximate the smallest t-interval required to complete the pattern. In other words, find the *period* of the pattern.

1. $x = 2\cos(t) + 2\cos(4t)$
   $y = 2\sin(t) + 2\sin(4t)$

2. $x = t - 2\sin(t)$
   $y = 2 - 2\cos(t)$

3. $x = 2\sin(2t)$
   $y = 2\sin(3t)$

4. $x = 2\cos(t) + 0.75\cos(\frac{4}{3}t)$
   $y = 2\cos(t) - 0.75\cos(\frac{4}{3}t)$

5. $x = t - \sin(t)$
   $y = 1 - \cos(t)$

6. $x = \cos(t) + 2\cos(0.5t)$
   $y = \sin(t) - 2\sin(0.5t)$

11.3: INVERSES, PARAMETRIC STYLE

The graph of any function $f(x)$ can be described parametrically by the equations

$$x = t, \quad y = f(t)$$

1. If the function $f(x)$ is one-to-one, then it has an inverse. In this case, the graph of $f^{-1}(x)$ can be described parametrically by the equations

   $$x = f(t), \quad y = t$$

   Explain why this is true.

2. Starting with the familiar function $f(x) = x^3$, show that the parametric equations $x = t$, $y = f(t) = t^3$ produce the same graph. (Use an interval for t that includes both negative and positive values.)

3. Show that the parametric equations $x = t^3$, $y = t$ produce the same graph as $f^{-1}(x)$, or $\sqrt[3]{x}$.

4. Let $g(x) = 3^x$. Write a set of parametric equations to describe the graph of $g(x)$. Write another set to describe the graph of $g^{-1}(x)$. Graph both sets of parametric equations on the same axes, using square scaling. What is the algebraic expression for $g^{-1}(x)$?

TI-82/83 GUIDE

This guide provides background on the TI-82/83 graphing calculators that will be useful for *Precalculus in Context: Projects for the Real World.* It consists of a basic tutorial followed by additional instructions relevant to each section in the laboratory manual. We do not attempt to show you everything that you can do on the TI-82/83.

BASIC TUTORIAL

As you proceed through this guide, specific keys that you are to press appear in a box. For example, you may be asked to press GRAPH. Operations corresponding to the colored lettering above the keys (blue or light gray on the TI-82, aqua or yellow on the TI-83) are indicated in brackets, [ ]. To access upper key functions, press 2nd or ALPHA, depending on the color of the lettering, and then the key.

Getting Started: On, Off, and Contrast

Turn the calculator on by pressing ON.

You may need to adjust the contrast. Press 2nd followed by holding down the up-arrow key ↑ to darken or the down-arrow key ↓ to lighten. (The four arrow keys are located directly below the TRACE and GRAPH keys.)

To turn your calculator off, press 2nd [OFF]. If you forget, the calculator will automatically turn off after a period of non-use.

Calculating and Editing

The screen that displays your calculations is called the home screen. Press CLEAR once or twice to begin with a clear home screen. You do *not* have to clear the screen after each computation.

Example: Compute 3×4.

After pressing 3 × 4, press ENTER. Note that the original problem, written as $3 * 4$, remains on the left side of the screen and the answer appears to the right.

Warning! *The TI-82/83 has two minus keys,* − *and* (-)*, to differentiate between the operation of subtraction (such as* $3 - 2 = 1$*) and the opposite of a number (such as* -2*). On your calculator screen, the subtraction sign appears slightly longer than the negative sign.*

Example: Compute $-2 + 5$.

To compute $-2 + 5$, press (-) (the key to the left of ENTER) then 2 to create the number -2. Finish the computation to get the answer 3.

Note: *If you press* − *instead of* (-) *in this problem, you get the wrong answer! Try it for yourself.*

Example: Compute $5 - 2$.

To compute $5 - 2$, press 5 − 2 ENTER.

Example: Compute 8^2 and 1.05^7.

Press [8] followed by [x^2] [ENTER]. You can also compute the square of eight by pressing [8] [∧] [2]. Now try 1.05^7 using [∧] [7] to compute the power.

Example: Compute $\sqrt{16}$.

Press [2nd] [[√]] (same key as [x^2]) followed by [1] [6] [ENTER]. (If you are using a TI-83, notice that a left parenthesis was automatically inserted after the square root sign. The calculator computed the square root even if you failed to add a right parenthesis after the 16.)

Example: Try to compute $\sqrt{-16}$.

You should have gotten an error message when you tried to compute $\sqrt{-16}$ because there is no real number whose square is -16. (If you are working on a TI-83 that is in (a + bi)-mode, then you would have gotten the answer $4i$. In that case, change the mode setting to Real and try this example again. See *Changing Mode Settings* on page 227.) To return to the home screen, press the number corresponding to Quit.

Example: Compute $\sqrt[5]{32}$.

Press [5] for the fifth root and then [MATH] [5] for $\sqrt[x]{}$. Next, press [3] [2] [ENTER]. Did you get 2?

Correcting an Error

We tackle two situations connected with making errors. First, we look at an example of an error that your calculator recognizes as an error. Then we provide an example that illustrates what you can do when you discover that you have punched in an error that the calculator is able to compute.

Correcting by Deleting

Example: Let's start by making a deliberate error: press [3] [+] [+] [2] [ENTER].

Your calculator will respond with an error message. Press the number corresponding to Goto and the cursor will direct you to the error. Erase one of the plus signs by pressing [DEL] for delete, and then press [ENTER]. The correct answer to 3 + 2 will appear.

Correcting by Inserting

Example: Press [3] [+] [4] [ENTER] and suppose that you really wanted $-33 + 4$.

Press [2nd] [[ENTRY]] to return to the previous command. Use the left-arrow key [←] to position the cursor over the 3. Press [2nd] [[INS]]. (Note the cursor changes from a box to a line.) Now press [(-)] [3] to insert -3 and then [ENTER] to complete the calculation.

Resetting the Memory

Warning! *Resetting the memory erases all data and programs. (Skip this step if you have programs or data that you do not wish to erase.) All calculator settings will return to the default settings including the contrast settings. So, you may need to adjust the contrast after resetting your calculator.*

Here's how to reset your calculator.

- Press [2nd] [[MEM]] to display one of the memory menus shown in Figure 1 or Figure 2.

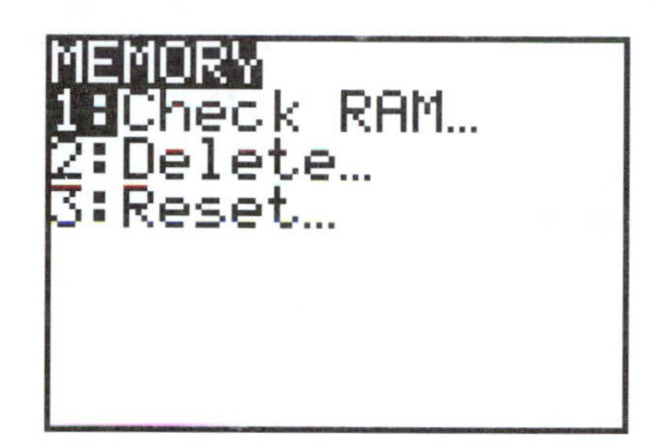

Figure 1 Memory Menu for TI-82

Figure 2 Memory Menu for TI-83

On the TI-82:

- Press [3] followed by [2] to reset the memory. This will clear all memory and return the calculator to its default settings. The message **Mem cleared** should appear on your screen.

On the TI-83:

- Press [5] and then [2] to return to the default settings without erasing the memory. (If you press [1], you will reset the memory and all programs, data, functions, and graphs will be erased.)

Changing Mode Settings

Press [MODE]. The cursor should be blinking on Normal. If you have reset your calculator as instructed above, the mode settings will match the default settings shown below.

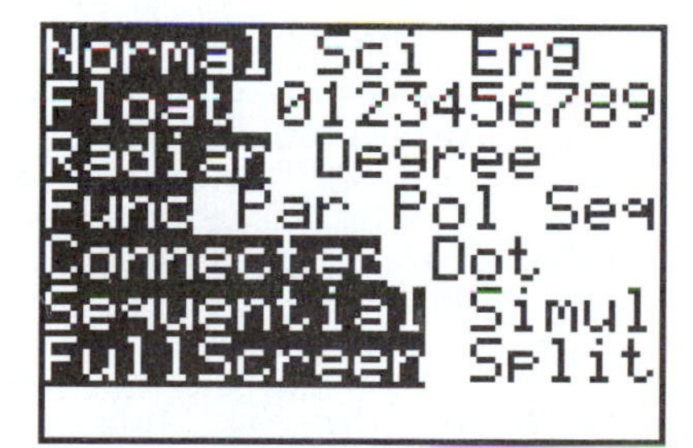

Figure 3 Mode Menu for TI-82

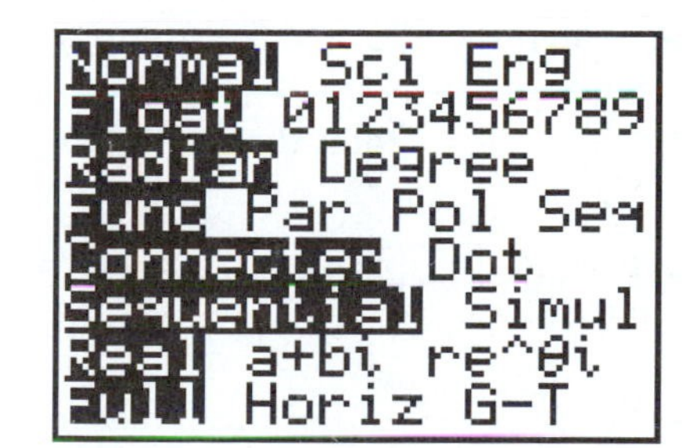

Figure 4 Mode Menu for TI-83

Suppose that you want all your answers to be displayed in scientific notation. Use the right-arrow key [→] to highlight Sci (for scientific notation) and then press [ENTER]. Press [2nd] [[QUIT]] to exit the mode menu.

Example: Compute 3654×1781.

The displayed answer should read: 6.507774 E6. (E is shorthand notation for $\times 10^6$.)

To return the mode setting to Normal, press MODE; the cursor should be blinking on Normal, so press ENTER. Press 2nd [QUIT] to exit the mode menu.

Graphing

First, check your mode settings: press MODE. For now, adjust the mode settings to match the default settings shown in Figure 3 or Figure 4.

Next, check to see if any functions have been stored in your calculator: press Y=. The screen in Figure 5, for example, indicates that two functions have been stored. Here's one way you could erase them.

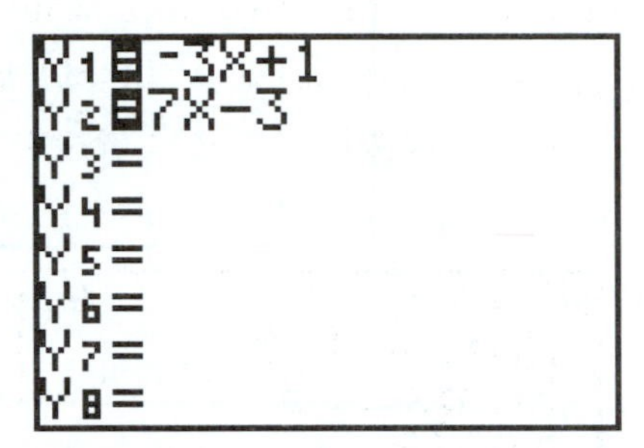

Figure 5 Y= Screen for TI-82

After pressing Y=, you should see a blinking cursor opposite Y1.

- Press CLEAR to erase the function stored in Y1.
- Then press ↓, to move the cursor opposite Y2, and again press CLEAR. If you need to clear more than two functions, continue this process.
- Press 2nd [QUIT] to exit the Y= screen.

If you did not have any functions stored in your calculator, test the preceding instructions after you have completed the next example. As you work through this guide, unless you are told otherwise, you should erase stored functions before you begin a new example.

Example: Graph $y = x^2$ and $y = x^3$ in the standard viewing window.

- Press Y= to enter the Y= screen. The blinking cursor should be to the right of Y1.
- Press the X,T,θ -key on the TI-82 or the X,T,θ,n -key on the TI-83 followed by x^2 to enter the first function.
- Press ↓ to position the cursor opposite Y2. To enter the second function, press X,T,θ or X,T,θ,n followed by ∧ 3.
- To graph $y = x^2$ and $y = x^3$ in the standard viewing window, press ZOOM 6.
- After viewing your graph, press WINDOW to check the settings of the standard viewing window. Your screen should match Figure 6 or Figure 7.

```
WINDOW FORMAT
 Xmin=-10
 Xmax=10
 Xscl=1
 Ymin=-10
 Ymax=10
 Yscl=1
```

Figure 6 Window Menu for TI-82

```
WINDOW
 Xmin=-10
 Xmax=10
 Xscl=1
 Ymin=-10
 Ymax=10
 Yscl=1
 Xres=1
```

Figure 7 Window Menu for TI-83

In the standard viewing window, the scaling on the x- and y-axes goes from -10 to 10 with tick marks one unit apart (since Xscl and Yscl = 1). At times we may refer to this window using the notation $[-10, 10] \times [-10, 10]$.

In the next example, you will continue where the previous example left off.

Example: Graph $y = x^2$ and $y = x^3$ in the window $[-1.5, 1.5] \times [-1.0, 2.0]$. Place tick marks every 0.5 units. (In this example, the window settings should match the ones on the screen in Figure 8.)

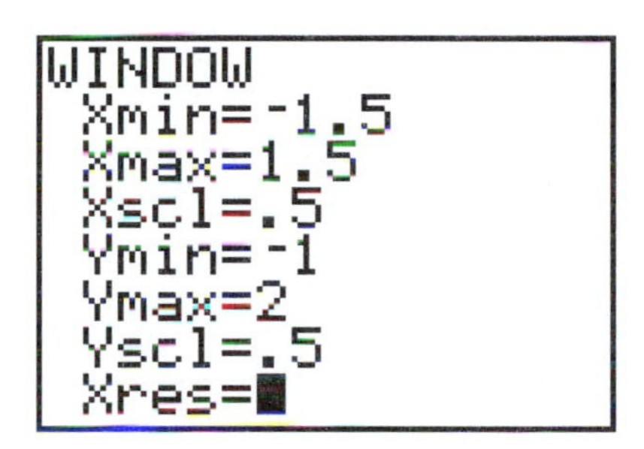

Figure 8 New Window Settings

- You should be looking at the window screen from the previous example. On the TI-82 it will be necessary to press [↓] to position the cursor to the right of Xmin. (The cursor will already be in this position if you are using a TI-83.)
- Press [(-)] [1] [.] [5] [ENTER] to change the value of Xmin.
- Now, your cursor should be to the right of Xmax. Change the value of Xmax to 1.5 and press [ENTER]. Continue changing the window settings until yours match the ones in Figure 8.
- Finally, press [GRAPH].

When you have completed this example, erase the functions in Y1 and Y2. (See the instructions at the beginning of *Graphing*.)

That's it! You have completed the basic tutorial. Now practice and experiment on your own with the calculator until you begin to feel comfortable with these basic operations. The remainder of this guide will introduce new techniques as they are needed, section by section, for your work in *Precalculus in Context: Projects for the Real World*.

Section-by-Section Guide

Note: *When entering a function that depends on x, the variable x is entered by pressing* [X,T,θ] *on the TI-82 and* [X,T,θ,n] *on the TI-83. These keys appear in the same location on both calculators. To avoid repetition we will ask you to press* [X,T,θ] *when we want you to press either of these keys.*

SECTION 1: LINEAR FUNCTIONS

Clearing Stored Functions

Unless you are told otherwise, erase any previously stored functions each time you begin a new example. The tutorial demonstrated one method for clearing functions. Here is another.

- Press [2nd] [[MEM]] [2] to get to the DELETE FROM menu.
- Press the number corresponding to Y-vars.
- Each time that you press [ENTER], you will delete the function to the right of the triangle marker. Press [ENTER] repeatedly until all functions have been cleared or use the arrow keys to selectively mark functions that you wish to clear.

Plotting Points

You can use your calculator to plot the Fahrenheit-Celsius data given in the preparation for Lab 1A. Then graph your guess for the formula that relates degrees Fahrenheit to degrees Celsius. This will allow you to check how closely the function specified by your formula follows the pattern of the data.

Example: Plot the data in the following table and then overlay the graph of $y = 18x + 85$.

Sample Data	
x	y
−2	40
−1	60
1	100
3	140

Step 1: Clear any stored functions. (See *Clearing Stored Functions.*)

Plan of action: The TI-82/83 has six lists, L1–L6. We'll store data from the x-column in L1 and the y-column in L2. However, before entering the data in these lists, we'll first erase any previously stored data. The instructions for erasing and entering the data are outlined in Steps 2 and 3. The directions for plotting the data are outlined in Step 4.

Step 2: Clear lists L1 and L2:

- Press [STAT] [4] [2nd] [[L1]] (the same key as [1]) [,] [2nd] [[L2]] and then [ENTER]. Your calculator will respond with the message Done.

Step 3: Enter the data as follows.

- Press [STAT] [1] for Edit. If necessary, use the left-arrow key [←] to highlight the cell directly under L1 as shown in Figure 9.

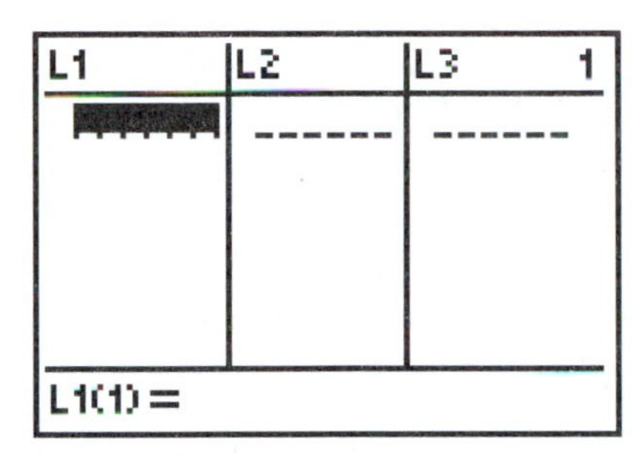

Figure 9 List Screen

- The screen above shows three empty lists, L1–L3. At the bottom of the screen, L1(1)= indicates that the first entry in L1 is highlighted. Press the right-arrow key [→] to highlight the first entry in L2 and observe that the bottom of the screen now displays L2(1)=. Press the left-arrow key [←] to return to L1(1).
- Enter the x-data in L1: press [(-)] [2] [ENTER] [(-)] [1] [ENTER] [1] [ENTER] [3] [ENTER].
- Press the right-arrow key [→] to move the cursor to the top of list L2. Enter the data from the y-column into L2.

Step 4: Plot the sample data.

- Press [2nd] [[STAT PLOT]] (same key as [Y=]). The menu that appears gives you a choice of three plots.
- Press [1] for Plot 1. Your screen should match Figure 10 or Figure 11.

Figure 10 TI-82 STAT PLOT

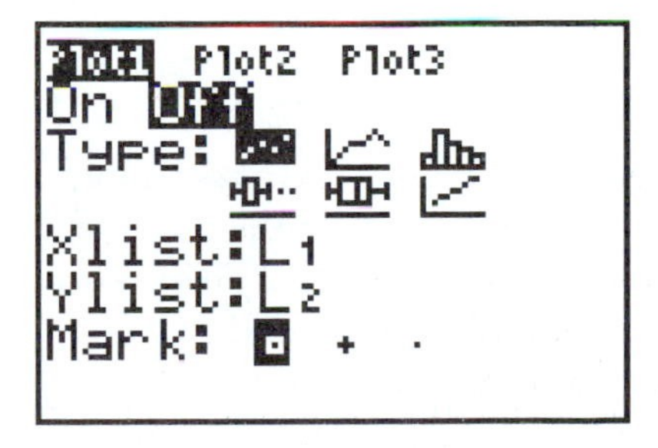

Figure 11 TI-83 STAT PLOT

- Press [ENTER] to highlight On.

- Press [↓] to move to Type. If not already highlighted, highlight the first choice (the scatter plot) and press [ENTER].
- Press [↓] to move to Xlist. On the TI-82, if L1 is not already highlighted, highlight L1, and then press [ENTER]. On the TI-83, press [2nd] [[L1]] if a different list is entered to the right of Xlist.
- If necessary, adjust the selection for Ylist. Then choose one of the three symbols following Mark.
- Press [WINDOW] and select appropriate settings for these data. For example, make sure that you choose a value for Xmin that is smaller than the x values in the table and a value for Xmax that is larger than the x values in the table. Now press [GRAPH].
- The ZOOM menu offers an option that is an alternative to setting the window manually and then graphing. Press [ZOOM], and then [↓] to scroll through the options. Now press [9] for ZoomStat. When you use this option, your calculator automatically selects a viewing window that displays all of your data.

Step 5: Add the graph of $y = 18x + 85$ to the plot of the sample data.

- Press [Y=]. Enter the function $y = 18x + 85$ as Y1. (Refer to *Graphing* beginning on page 228.)
- Press [GRAPH]. (Your display should show four data points that lie fairly close to a line.)

Warning! *If you fail to turn off STAT PLOT 1, this plot will appear superimposed on your next graph. Worse, if you subsequently erase the data in L1 and L2 without turning off this plot, you will get error messages each time you press* [GRAPH] *because your calculator will still be trying to create STAT PLOT 1.*

Step 6: Turn off all Plots as follows.

- Press [2nd] [[STAT PLOT]] [4] for PlotsOff.
- Press [ENTER]. Your calculator should respond with the message **Done**.

Adjusting the Viewing Window for Square Scaling

The viewing screen on your calculator is a rectangle. Therefore if you use the standard window, the tick marks on the y-axis will be closer together than those on the x-axis. For square scaling, we want the distance between 0 and 1 on the x-axis to be the same as the distance between 0 and 1 on the y-axis.

Example: Graph the line $y = x$ in the standard window and then switch to square scaling.

Step 1: Graph $y = x$ in the standard viewing window.

- Enter $y = x$ as Y1.

- Press ZOOM 6 for ZStandard.

Observe the spacing between the tick marks on the x- and y-axes. Notice that the tick marks on the y-axis are closer together than the tick marks on the x-axis and that the graph does not make a 45° angle with the x-axis.

Step 2: Change to a square viewing window as follows:

- Press ZOOM 5 to select ZSquare. When the graph appears on your screen, observe the equal distance between tick marks on the two axes. Notice that the graph makes a 45° angle with the x-axis.
- Press WINDOW and note the changes in the viewing window settings.

SECTION 2: QUADRATIC FUNCTIONS

Graphing a Quadratic Function

When you graph a quadratic function, it is important to experiment with various viewing windows to ensure that you have captured all the important features of the graph on your screen.

Example: Graph $y = 2x^2 - 2x + 12$ using the three different viewing windows specified in the directions below.

Start by entering the function.

- Press Y= and enter $2x^2 - 2x + 12$. (Remember to use − when you subtract $2x$.)
- Now press WINDOW.

Window 1: Adjust the settings to match those in Figure 12 and then press GRAPH. Your graph should be the familiar U-shape of a parabola.

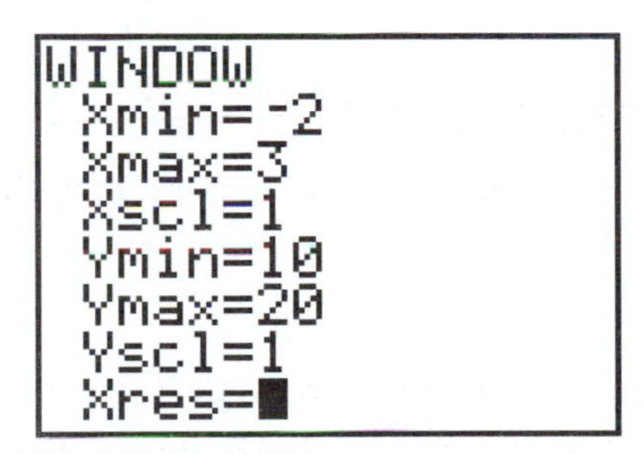

Figure 12 Settings for Window 1

Next, we view the graph of this quadratic function in two other windows. In each case, the viewing window selected fails to show key features of the parabola.

Window 2: Change the Xmin setting to 2 as follows.

- Press [WINDOW]. (If you are using a TI-82, remember to press the down arrow key [↓] to move the cursor opposite Xmin.) Change the value for Xmin to 2.
- Now, press [GRAPH]. Notice that in Window 2 the graph of $y = 2x^2 - 2x + 12$ looks more like a line than a parabola.

Window 3: Change to the standard viewing window by pressing [ZOOM] [6]. What do you see? In general, if you ask the calculator to graph a function, and just see empty axes, you probably are looking at a part of the plane that contains none of the graph.

Calculating Outputs of a Function

Example: Find the output of $f(x) = x^2 - x + 6$ when x has value 4.

- Press [Y=], erase any previously stored functions, and enter $x^2 - x + 6$ as Y1.
- Press [2nd] [[QUIT]] to return to the home screen. Then:

  On the TI-82: press [2nd] [[Y-VARS]] [1] to select Function.

  On the TI-83: press [VARS] [→] to highlight Y-VARS and then [1] to select Function.

- Press [1] to select Y1.
- Next, in parentheses, enter the value for which you want the output: press [(] [4] [)] and then [ENTER]. The output should be 18.

Now, use the method outlined above to check that $f(-2) = 12$.

Using TRACE and ZBox to Approximate the Coordinates of a Point on the Graph

Example: Approximate the vertex (turning point) and x-intercepts of the parabola $y = 2x^2 + 3x - 2$.

Plan of action: First, we'll graph $y = 2x^2 + 3x - 2$ in the standard viewing window. Then we'll use ZBox to get a window that magnifies the section of the graph containing the vertex and x-intercepts. Finally, we'll use TRACE to estimate the vertex and x-intercepts.

Step 1: Enter and graph $y = 2x^2 + 3x - 2$ in the standard viewing window.

- Enter the function $2x^2 + 3x - 2$ as Y1.
- Press [ZOOM] [6] to graph this function in the standard viewing window. Your graph should resemble the one in Figure 13 (without the box).

Step 2: Magnify the section of graph containing the vertex and x-intercepts by drawing a box that will become the new viewing window.

- Press [ZOOM] [1] for ZBox. There should be a blinking cursor (a plus sign with a blinking pixel in the middle) at the center of your screen. (You may have trouble seeing the cursor until you move it off the origin.)

- Next, we are going to draw a box like the one in Figure 13. Start by pressing the left ← and up ↑ arrow keys to move the cursor to the location where we want the upper left corner. Then press ENTER.
- Now, use the right → and down ↓ arrow keys to draw a box similar to the one shown in Figure 13. Press ENTER to replace the window settings with those that correspond to the box.

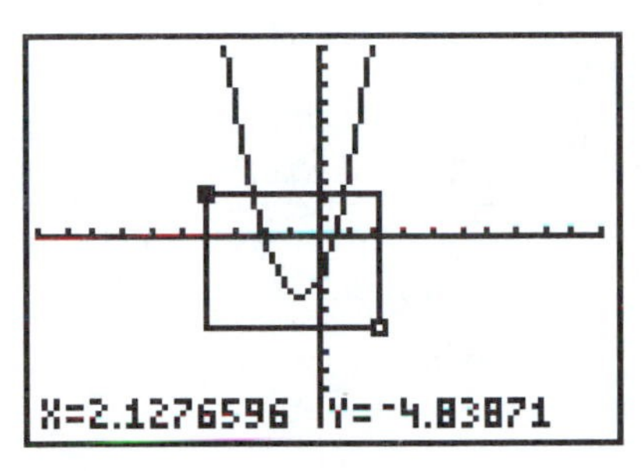

Figure 13 Graph of $y = 2x^2 + 3x - 2$

Step 3: Use TRACE to find the approximate coordinates of the x-intercepts and vertex.

- Press TRACE. Use the right → and left ← arrow keys to position the cursor at the vertex of the parabola. Then read the approximate coordinates of the vertex at the bottom of your screen.
- Continue to use the right → and left ← arrow keys to position the cursor on each of the x-intercepts. Read the approximate coordinates for each of the x-intercepts.

Your answers should be similar to those in the screens in Figure 14.

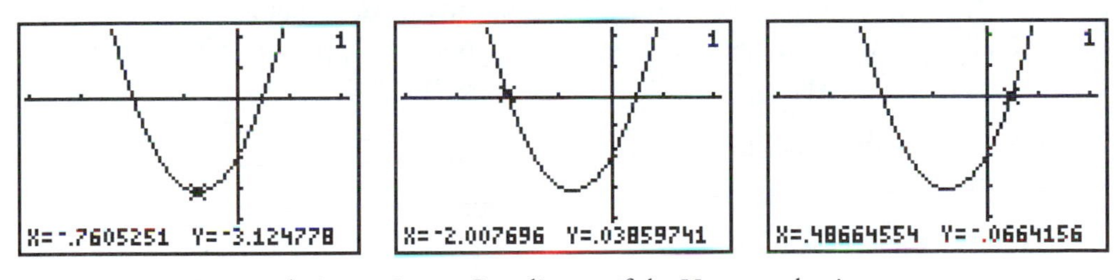

Figure 14 Approximate Coordinates of the Vertex and x-intercepts

SECTION 3: SHIFTING AND REFLECTING GRAPHS

In Lab 3, *Graph Trek Explorations,* you will be investigating the effect that certain algebraic modifications, such as adding a constant to the input variable, have on the graph of a function. You'll want to experiment using several different functions. We've provided some functions and algebraic modifications that you might want to consider.

Using Parentheses

Warning! *When you want to apply a function to an expression, you must enclose the entire expression in parentheses. For the built-in functions (such as the square root or the sine function), the TI-83 will automatically insert the left parenthesis and you will need only add the right parenthesis. However, this is not the case on the TI-82 and you will need to enter both the left and right parentheses.*

Example: Graph $y = \sqrt{x+2}$ in the window $[-5, 5] \times [-5.5]$.

- Enter $y = \sqrt{x+2}$ as follows:

  On the TI-82: press [Y=] [2nd] [[√]] [(] [X,T,θ] [+] [2] [)].

  On the TI-83: press [Y=] [2nd] [[√]] [X,T,θ,*n*] [+] [2] [)]. (Note: the TI-83 automatically inserts the left parenthesis when you press the square root key.)
- Graph this function in the window $[-5, 5] \times [-5.5]$. Your graph should resemble the one in Figure 15.

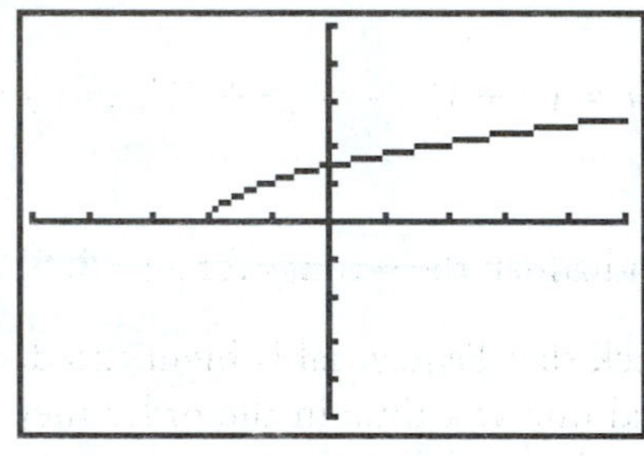

Figure 15 Graph of $y = \sqrt{x+2}$

Graphing Trigonometric Functions in the Trig Viewing Window

Locate the [SIN], [COS], and [TAN]-keys on your calculator. These keys will allow you to study the graphs of the sine, cosine, and tangent functions before the functions are formally introduced. Before entering any of these functions, press [MODE]. If Radian is not highlighted, move the cursor to Radian and press [ENTER].

Example: Examine the graphs of $y = \sin(x)$ and $y = \sin(x + 2)$ in the trig viewing window.

- Enter the function $\sin(x)$ as Y1:

  On the TI-82: press [SIN] [(] [X,T,θ] [)].

  On the TI-83: press [SIN] [X,T,θ,*n*] [)].
- Next, enter the function $\sin(x + 2)$ as Y2:

  On the TI-82: press [SIN] [(] [X,T,θ] [+] [2] [)]. On the TI-83: use the same set of keystrokes but omit [(].
- Press [ZOOM] [7] for the trig viewing window. You should see two wavy curves. (If you don't, press [MODE]. You have probably forgotten to highlight Radian.)
- Finally, press [WINDOW] and observe the settings for the trigonometric window.

Using the Absolute Value Function

Example: Graph $y = |x|$ in the standard viewing window.

- Enter $y = |x|$ as Y1 as follows:

  On the TI-82: press [Y=] [2nd] [[ABS]] (same key as [x^{-1}]) [(] [X,T,θ] [)].

  On the TI-83: press [Y=] [2nd] [CATALOG]. (The triangle marker should already be pointing to abs.) Press [ENTER] followed by [X,T,θ,n] [)].

- Press [ZOOM] [6] to view the V-shaped graph of the absolute value function.

Graphing a Family of Functions

Using your calculator's list capabilities, you can substitute each value in a given list for a constant in an algebraic formula. This feature allows you to graph an entire family of functions quickly. On the TI-82/83, you specify a list by enclosing the members of the list in brackets: { }.

Example: Graph the family $y = (x + 1)^2, y = (x + 2)^2$, and $y = (x + 3)^2$ in the window $[-5, 5] \times [-1, 10]$.

- Press [WINDOW] and adjust the settings for a $[-5, 5] \times [-1, 10]$ window.
- Press [MODE]. Check that Sequential is highlighted. (Using this mode setting, functions are graphed one at a time in the order they have been entered.) Press [2nd] [[QUIT]] to exit the mode menu.
- Enter the three functions by specifying the constants, 1, 2, and 3, in a list as follows. Press [Y=] [(] [X,T,θ] [+] [2nd] [[{]] [1] [,] [2] [,] [3] [2nd] [[}]] [)] [x^2]. Then press [GRAPH] and watch as the three functions are graphed one after the other.

SECTION 4: EQUATIONS AND INEQUALITIES

Finding the Points of Intersection of Graphs

Example: Graph $f(x) = -2x^2 + 2x + 12$ and $g(x) = -4x + 15$ and determine the points where the graphs intersect.

You could approximate the solutions to this problem using the techniques outlined in the example on page 234. However, the TI-82/83's intersect command will provide, when possible, exact solutions and, if not possible, very good approximations.

Step 1: Graph both functions in a viewing window that gives a clear view of the points of intersection.

- Enter the two functions into your calculator.
- Adjust the WINDOW settings so that you can see both points of intersection. (*Hint:* you might start with the standard viewing window, [ZOOM] [6], and then adjust the WINDOW settings after looking at the graph.)

Step 2: Approximate the coordinates of one of the points of intersection.

- Press 2nd [CALC] (the key above TRACE) and then press 5 for intersect. The image on your screen should be similar the one shown in Figure 16.

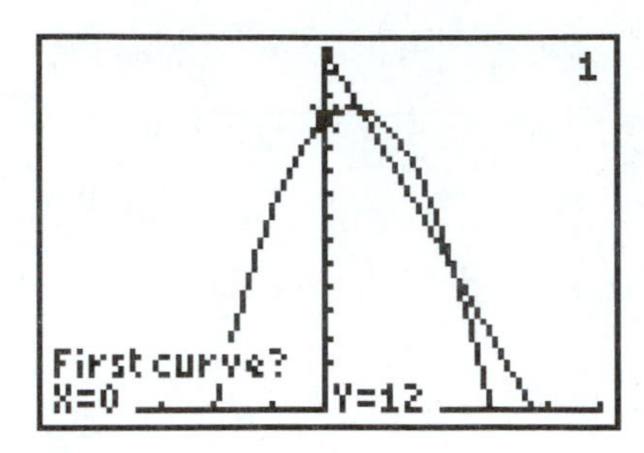

Figure 16 The Intersections of Two Graphs

- Try pressing the up ↑ and down ↓ arrow keys. The cursor will jump back and forth from the line to the parabola. Position the cursor on the line (we'll designate the line to be the first curve) and press ENTER. Then (if necessary) position the cursor on the parabola (the second curve), and again press ENTER.
- Now, you must provide your calculator with a guess for the point of intersection. Press → or ← to move the cursor close to one of the points of intersection. Press ENTER. Read off the (approximate) coordinates for this point of intersection.

Step 3: Find the coordinates of the second point of intersection by repeating the process outlined in Step 2.

If you have done everything correctly, you will find that the two graphs intersect at approximately (0.634, 12.464) and (2.366, 5.536).

Graphing Piecewise Defined Functions

Example: Graph the piecewise defined function $f(x) = \begin{cases} x - 4 \text{ if } x > 4 \\ -x + 4 \text{ if } x \leq 4 \end{cases}$

The graph of $f(x)$ consists of two half-lines pieced together. You'll want the graph of $y = x - 4$ when x-values are greater than 4 and $y = -x + 4$ when x-values are less than or equal to 4.

Step 1: Enter the functions that you want to piece together.

- Press Y=. Enter $y = x - 4$ as Y1 and $y = -x + 4$ as Y2.
- Press ZOOM 6. Your graph should look like a cross: ×.

Step 2: Form the function $f(x)$. This is where you'll piece together the graphs of Y1 and Y2.

- Press Y= and ↓ to move the cursor opposite Y3.

Here's how to enter $f(x)$ as Y3:

- To insert Y1:

  On the TI-82: press [2nd] [[Y-VARS]] [1] [1].

  On the TI-83: press [VARS] [→] to highlight Y-VARS and then press [1] [1].

- Press [×]. Next, enter the condition that governs when to use Y1: press [(] [X,T,θ] [2nd] [[TEST]] [3] [4] [)].
- Press [+].
- To insert Y2:

  On the TI-82: press [2nd] [[Y-VARS]] [1] [2].

  On the TI-83: press [VARS] [→] [1] [2].

- Press [×]. Then enter the condition that governs when to use Y2: press [(] [X,T,θ] [2nd] [[TEST]] [6] [4] [)]. When you have completed this step, your screen should be similar to Figure 17.

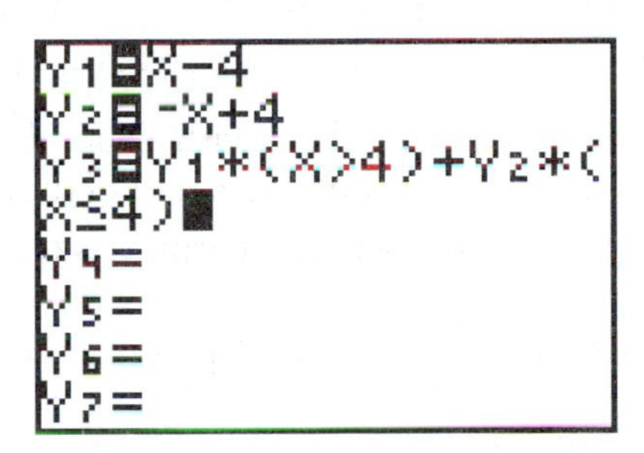

Figure 17 Y= Screen on TI-82

Here's how your calculator interprets the information you've just entered as a piecewise defined function. The calculator assigns the expression $(x > 4)$ the value 1 when the inequality is true (in other words, when the input variable, x, is greater than 4). In this case, the inequality $x \leq 4$ is false, so the calculator sets the expression $x \leq 4$ equal to 0. Thus, for $x > 4$, the function Y3 is equivalent to:

$$\text{Y3} = (x - 4)(1) + (-x + 4)(0) = x - 4.$$

And when $x \leq 4$ the function Y3 is equivalent to:

$$\text{Y3} = (x - 4)(0) + (-x + 4)(1) = -x + 4.$$

Step 3: Graph $f(x)$.

- Turn off Y1 and Y2: Position the cursor over the equals sign opposite Y1. Press [ENTER]. Then, position the cursor over the equals sign opposite Y2 and press [ENTER].
- Now, press [GRAPH]. The graph of $f(x)$ should look V-shaped.

Fitting a Line to Data

If your data, when plotted, lie exactly on a line, you can use algebra to determine the equation of the line. However, real data seldom fall precisely on a line. Instead, the plotted data may exhibit a roughly linear pattern. The least squares line (also called the regression line) is a line that statisticians frequently use when describing a linear trend in data.

Example: Use the least squares line to describe the linear pattern in the data below.

x	y
−3.0	−6.3
−2.0	−2.8
1.2	2.0
2.0	4.1
3.1	5.0
4.2	7.2

Step 1: Enter the data from the table above into lists L1 and L2 and make a scatter plot of the data as follows.

- Erase any stored functions or any stored data in lists L1 and L2. Then enter the data from the x-list into L1 and those from the y-list into L2. (Refer to *Plotting Points* on pages 230–232.)
- After entering the data, press [2nd] [[STAT PLOT]]. Press [ENTER] twice to select Plot 1 and turn it ON. Make sure that the first choice to the right of TYPE (the scatter plot) is highlighted. If necessary, select L1 for Xlist and L2 for Ylist.
- Press [ZOOM] [9] to plot the data in the ZoomStat viewing window.

Note: *ZoomStat adjusts the WINDOW settings so that all data points are visible on the graphing screen.*

Step 2: Next, we'll find the equation for the least squares line.

On the TI-82:

- Press [STAT], use [→] to highlight CALC, and then press [3] for SetUp. Check to see that L1 is highlighted to the right of Xlist under 2-Var Stats. If not, select L1 by pressing [↓] until L1 (opposite Xlist) is highlighted and press [ENTER]. If necessary, select L2 for Ylist.
- Again, press [STAT], use [→] to highlight CALC, and then press [5] for LinReg(ax+b), [ENTER]. The output will tell you the values for a and b, the slope and y-intercept, respectively.

On the TI-83:

- Press [STAT], use [→] to highlight CALC, and then press [4] for LinReg(ax+b) [ENTER]. Then press [2nd] [[L1]] [,] [2nd] [L2] [ENTER]. The output will tell you the values for a and b, the slope and y-intercept, respectively.

Note: *Statisticians frequently specify the form of this line as* y = a + bx, *where* a *and* b *represent the* y*-intercept and slope, respectively. In fact, the STAT CALC menu provides a selection that corresponds to statistical convention: select* [9] *on the TI-82 and* [8] *on the TI-83.*

Step 3: Graph the least squares line and the data in the same viewing window.

- Press [Y=]. The cursor should be blinking to the right of Y1. Insert the equation for the least squares line as follows: press [VARS] [5] for Statistics, use the right-arrow key [→] to highlight EQ, and then press the number corresponding to RegEq ([7] on the TI-82 and [1] on the TI-83).
- Now, press [GRAPH]. The picture on your calculator's screen should be similar to Figure 18.

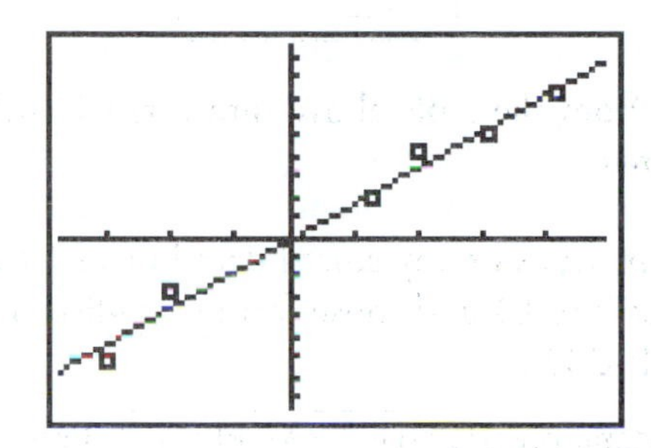

Figure 18 Scatter Plot and Regression Line

Warning! *Remember to press* [2nd] [[STAT PLOT]] [4] [ENTER] *to turn off all plots.*

SECTION 5: POLYNOMIAL FUNCTIONS

Determining the Roots of a Function

Determining the roots (zeros, or x-intercepts) of a function using factoring often requires considerable skill; and most polynomials can't be factored easily, if at all. However, your calculator can take much of the drudgery out of finding roots.

Example: Find the roots of the polynomial function $g(x) = \frac{1}{2}x^4 + 2x^3 + x^2 - 3x - 5$.

Step 1: Graph $g(x)$ in the standard viewing window: enter $g(x)$ as Y1 and press [ZOOM] [6] to view the graph.

Step 2: Let's approximate the negative root first. You could zoom in on this intercept and then use TRACE to estimate the x-coordinate. However, here is another way to determine the root.

- Press [2nd] [[CALC]] [2] for root or zero.
- Select a lower bound: use [←] to move the cursor along the graph until it lies to the left of the negative root and then press [ENTER].

- Select an upper bound: use [→] to move the cursor to the right of the negative root (*but still to the left of the positive root*) and then press [ENTER].
- Guess the root: use [←] and/or [→] to move the cursor as close to the negative root (left most x-intercept) as possible and then press [ENTER].
- Read off the approximate value of the root (or zero) from the bottom of your viewing screen. The x-value should be close to -3.05 and the corresponding y-value very close to zero. (Note that $-5E-13$ is $-.0000000000005$.)

Step 3: Next, approximate the positive root: press [2nd] [[CALC]] [2] and then adapt the instructions in Step 2. If you have adapted the instructions correctly, you should get approximately 1.39 for this root.

Finding Local Maxima or Minima of Functions

In Section 2, you found the vertex of a parabola using a procedure requiring a combination of [ZOOM] [ZBox] and [TRACE]. (See page 234 for details.) The coordinates of any function's turning points could be estimated using this method. One drawback to this procedure is that you must frequently apply it several times in succession before you can obtain the desired accuracy. Here is another method for approximating local maxima or minima of a function.

Example: Let's estimate the local maximum and minimum of the cubic function $f(x) = x^3 - 4x^2 + 2x - 4$.

Step 1: Graph $f(x)$ using a viewing window that gives you a clear view of the two turning points (one peak and one valley) of the graph.

Step 2: Approximate the coordinates of the turning point associated with the local maximum (y-coordinate of the peak on the graph) as follows.

- Press [2nd] [[CALC]] and then press [4] for maximum.

Next, you need to specify an x-interval that contains the x-coordinate associated with the local maximum of the function:

- Specify a lower bound for this interval. Use [←] to select an x-value that is less than the x-value of the turning point of the graph's peak. Then press [ENTER]. A black triangle that points to the right will mark this lower bound.
- Specify an upper bound for this interval. Press [→] to specify an x-value that is larger than the x-value of the turning point of the graph's peak. Then press [ENTER]. A black triangle that points to the left will mark this upper bound.
- Provide a guess for the x-value corresponding to the local maximum. Press [←] and/or [→] to position the cursor at the top of the turning point and then press [ENTER]. The coordinates of this turning point will appear at the bottom of your screen. Compare your answer with the one shown in Figure 19.

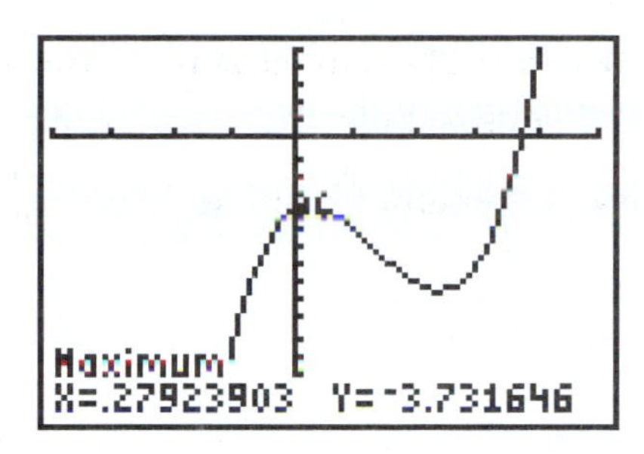

Figure 19 Approximation of Local Maximum

Step 3: Approximate the coordinates of the turning point associated with the local minimum (y-coordinate of the valley on the graph).

- Press [2nd] [[CALC]] and then press [3] for minimum.
- To determine the value of the local minimum, follow the last three instructions of Step 2: specify a lower bound and an upper bound containing the turning point (the valley), provide a guess for this turning point, and then read off the coordinates at the bottom of your screen. You should get a value for y that is close to -8.42.

SECTION 6: RATIONAL FUNCTIONS

Graphing a Rational Function

Warning! *If the numerator or the denominator of a rational function consists of more than one term, you must enclose it in parentheses when you enter it into your calculator.*

Example: Graph $r(x) = \dfrac{x^2 - 1}{x - 3}$.

- Press [Y=] and enter $r(x)$: press [(] [X,T,θ] [x^2] [−] [1] [)] [÷] [(] [X,T,θ] [−] [3] [)].
- Press [ZOOM] [6] to view the graph in the standard viewing window. Your graph should resemble the one in Figure 20.

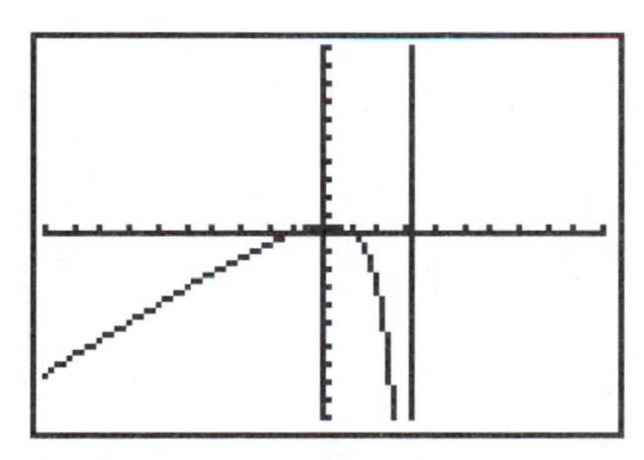

Figure 20 Graph of $r(x)$ in Standard Viewing Window

The vertical line in the graph above indicates that this function has a vertical asymptote at $x = 3$. Remember, this line is not part of the graph of the function. Furthermore, because the domain of this function is all real numbers except $x = 3$, there is a branch of this graph that lies to the right of the line $x = 3$. In order to observe this branch, you would have to adjust the Window setting for Ymax.

Do *not* clear this function from your calculator's memory until after you have completed the next topic, *Zooming Out.*

Zooming Out

Example: Graph the function $r(x) = \dfrac{x^2 - 1}{x - 3}$ in the standard viewing window and then zoom out by a factor of 4 several times.

For this example, observe what happens to the appearance of the graph of $r(x)$ as we "back away" by increasing the width and height of the viewing window. You should have the graph of $r(x)$ (from the previous example) in your screen.

- Press [WINDOW] and set Xscl and Yscl to 0. (This turns off the tick marks that appear on the axes. If you skip this step, the axis will get crowded with tick marks when you zoom out.)
- Check the settings for the zoom factor: press [ZOOM] and highlight MEMORY. Select [4] for SetFactors. If necessary, adjust the factor settings so that XFact = 4 and YFact = 4.
- To view the graph over wider x- and y-intervals, first press [ZOOM] [3] for Zoom Out. (A blinking pixel, part of a free moving cursor, should appear in the center of your screen. The zooming will be centered around this location. If you wish to change the focal point of the zooming, use the arrow keys to position this cursor at a different center.) Now, press [ENTER] to view the graph over wider x- and y-intervals.
- Press [ENTER] again to zoom out a second time. (The graph should look like a line except, perhaps, for a small blip slightly to the right of the origin.)
- Press [WINDOW] to observe the effect on the WINDOW settings of twice zooming out by a factor of 4. (The default setting for Zoom Out widens both the x- and y-intervals by a factor of each time that it is applied. In the previous example, you zoomed out twice. Therefore, the x- and y-intervals are 16 times wider than they were before you zoomed out.)

In the next example we'll change the default zoom settings in order to observe the behavior of a function that begins to act like its horizontal asymptote.

Example: Graph $q(x) = \dfrac{5x^2 + 20x - 105}{2x^2 + 2x - 60}$ in the standard viewing window.

If you have entered the function correctly, your graph should look like Figure 21.

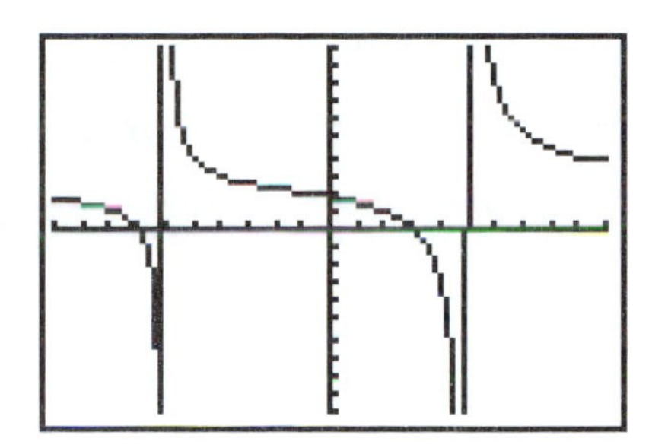

Figure 21 Graph of $q(x)$

Based on this graph you may suspect that the function $q(x)$ has a horizontal asymptote but this is not at all obvious. Let's observe the function's graph over increasingly wide x-intervals to see if it begins to behave like a horizontal line. Instructions on how to zoom out in just the horizontal direction follow.

- Press WINDOW and turn off the tick markings by setting Xscl and Yscl to 0.
- Press ZOOM and highlight MEMORY. Select 4 for SetFactors. To leave the y-interval unchanged and zoom out in the horizontal direction only, set YFact to 1 and XFact to 4.
- Press GRAPH to return to the graph of $q(x)$. Now press ZOOM 3 ENTER to widen the x-interval by a factor of four. Press ENTER several more times to continue widening the x-interval. Your graph should begin to resemble its horizontal asymptote $y = 2.5$.

SECTION 7: EXPONENTIAL GROWTH AND DECAY

Graphing Exponential Functions

Because exponential functions increase or decrease very quickly in certain regions of their domain, you may have to experiment in order to find a viewing window that captures the function's important graphical features.

Example: Graph $y - 4^x$ in the standard viewing window. Then change the window to $[-2, 2] \times [0, 10]$.

- Enter $y = 4^x$ by pressing Y= and then 4 ^ X,T,θ.
- Press ZOOM 6 to graph this function in the standard viewing window.

Notice that, to the left of the y-axis, the graph appears to merge with the line $y = 0$ and, to the right of the y-axis, the graph becomes so steep that it appears vertical.

- Press WINDOW. Now change the settings to the window $[-2, 2] \times [0, 10]$ so that the graph will fill more of the screen. Your graph should resemble the one in Figure 22.

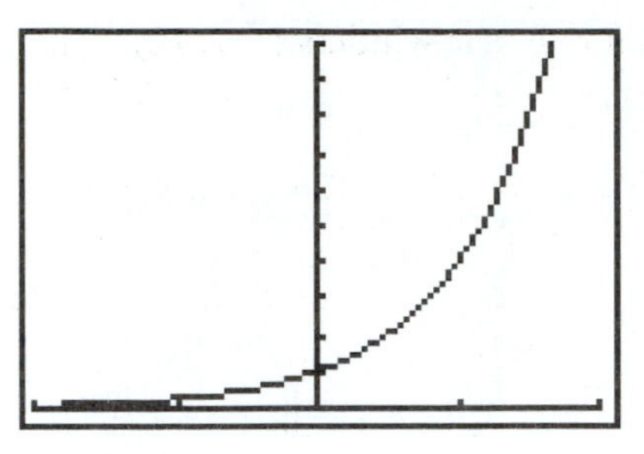

Figure 22 Graph of $y = 4^x$

There are two bases for exponential functions, 10 and e, that are so common they have their own function keys on the calculator [$[10^x]$] and [$[e^x]$].

Example: Graph $f(x) = e^x$ in the window $[-2, 3] \times [-1, 12]$.

- To enter $f(x)$:

  On the TI-82: press [Y=] [2nd] [$[e^x]$] (same key as [LN]) [(] [X,T,θ] [)].

  On the TI-83: press [Y=] [2nd] [$[e^x]$] (same key as [LN]) [X,T,θ,n] [)]. (Remember the TI-83 automatically inserts the left parenthesis.)

- Press [WINDOW] and adjust the settings for the window $[-2, 3] \times [-1, 12]$. Then press [GRAPH] to view the graph. The graph should have the same basic shape as the graph of $y = 4^x$.

Example: Graph $y = e^{-x/2}$ in the standard viewing window

There are two things that you must remember when entering this function. Use the [(-)] key for the opposite of $\frac{x}{2}$ and enclose the exponent, $-\frac{x}{2}$, in parentheses. The shape of your graph should resemble the graph in Figure 22 reflected over the y-axis.

Turning Off the Axes

Sometimes it is helpful to view a graph without the presence of the x- and y-axes. Here's how to turn off the axes.

On the TI-82: press [WINDOW], [→] for FORMAT, highlight AxesOff, and press [ENTER].

On the TI-83: press [2nd] [[FORMAT]], highlight AxesOff, and press [ENTER].

Try graphing one of the functions in the examples above with the axes turned off. (Remember to turn the axes back on when you are finished experimenting.)

SECTION 8: LOGARITHMIC FUNCTIONS

Graphing Logarithmic Functions

The logarithmic functions with base e and base 10 have their own function keys [LN] and [LOG], respectively. Logarithmic functions of other bases can be graphed by dividing these functions by the appropriate scaling factor.

Example: Graph $h(x) = \ln(x)$ in the window $[-1, 12] \times [-2, 3]$.

- To enter $h(x)$:

  On the TI-82: press [Y=] [LN] [(] [X,T,θ] [)].

  On the TI-83: press [Y=] [LN] [X,T,θ] [)].

- Press [WINDOW] and adjust the settings for the window $[-1, 12] \times [-2, 3]$. Then press [GRAPH] to view the graph.

SECTION 9: TRIGONOMETRIC FUNCTIONS

Graphing Trigonometric Functions

Three of the six basic trigonometric functions are built-in functions on the TI-82/83: sine [SIN], cosine [COS], and tangent [TAN]. Before graphing any of these functions, you should first check that your calculator is set in radian mode. Press [MODE] and highlight Radian if it is not already highlighted.

Example: Graph $y = \sin(x)$ and $y = \csc(x)$ in the trigonometric viewing window.

- Enter $y = \sin(x)$:

  On the TI-82: press [Y=] [SIN] [(] [X,T,θ] [)].

  On the TI-83: press [Y=] [SIN] [X,T,θ] [)]. (Remember that the TI-83 automatically inserts the left parenthesis.)

- Enter $y = \csc(x)$:

  Note: *Since* $\csc(x)$ *is defined as* $\dfrac{1}{\sin(x)}$, *we will use the reciprocal key,* [x^{-1}].

  Warning! *The* [SIN^{-1}]*-key on your calculator is NOT the same as the csc function.*

  On the TI-82: press [(] [SIN] [(] [X,T,θ] [)] [)] [x^{-1}].

  On the TI-83: press [(] [SIN] [X,T,θ,n] [)] [)] [x^{-1}].

- To graph these functions in the trig viewing window, press [ZOOM] [7]. Your graph should be similar to Figure 23.

Note: *The vertical lines indicate vertical asymptotes and not part of the graph of* $y = \csc(x)$

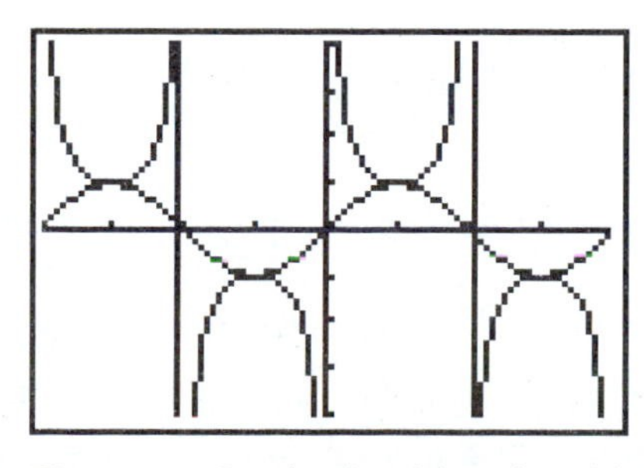

Figure 23 Graph of $\sin(x)$ and $\csc(x)$

The trigonometric viewing window gives a good picture of $\sin(x)$ and $\csc(x)$. Keep in mind, however, that it is not the best window for viewing all trigonometric functions. For example, it would not be a good viewing window for the function $y = 5\cos(10x)$. For this function, you would need to adjust the window settings in order to display the key features of its graph.

Doing Away with Dull Trig Tables!

With the TI-82/83 you can solve problems in right-triangle trigonometry without using trig tables. To compute the sine, cosine, or tangent of an angle measured in degrees, first change the Radian/Degree mode setting to Degree: press [MODE] and highlight Degree by moving the cursor over Degree and pressing [ENTER].

Example: Compute $\sin 30°$.

- Press [SIN] [3] [0] [ENTER]. (If you do not get 0.5 for the answer, go back and check that you have changed your calculator to degree mode.)

Note: *The TI-83 will forgive you if you fail to press* [)] *after entering* sin(30.

Example: Compute $\cos^{-1}(0.5)$ (recall this is the angle whose cosine is 0.5) and $\tan^{-1}(2.5)$.

- Press [2nd] [[COS^{-1}]] [.] [5] [ENTER]. Did you get 60? Remember this is 60° if you're in degree mode.
- Check that $\tan^{-1}(2.5) \approx 68.199°$.

Finally, here's a way to compute the sine, cosine, or tangent of an angle given in degrees without changing the mode setting. First, return your calculator to its default mode setting, Radian: press [MODE], highlight Radian, and press [ENTER]. Now, let's compute cos (60°) and sin (35°) without changing the mode setting. (This is a good idea because for most of your work in precalculus, you want radian mode.)

Example: Compute $\cos 60°$ and $\sin 35°$.

- Press [COS] [6] [0] [2nd] [[ANGLE]] [1] [ENTER]. You should get 0.5 for your answer.
- Now use this method for sin(35°). Did you get approximately 0.574?

SECTION 10: MULTIVARIABLE FUNCTIONS

Graphing a Function with More Than One Input Variable

You can represent functions with more than one input variable graphically by replacing one (or more) of the independent variables with a list of values. (If you don't remember how to handle lists, refer to *Graphing a Family of Functions* on page 237.)

Example: Examine the behavior of $F(w, x) = 3w - 2x$ by holding w constant, first at -2, then at 0, and then at 2, while x varies.

The three graphs, taken together, show how F varies with x for three different values of w. By forming a list $\{-2, 0, 2\}$ of the constant values for w, you can produce the graphs $y = F(-2, x)$, $y = F(0, x)$, and $y = F(2, x)$ using a single functional expression. Here's how:

- Press [Y=] and then [3] [2nd] [[{] [(-)] [2] [,] [0] [,] [2] [2nd] [[}] [−] [2] [X,T,θ]. Notice that the input variable w has been replaced by the list of values $\{-2, 0, 2\}$.
- Press [ZOOM] [6] and watch as three parallel lines are graphed one by one.

SECTION 11: PARAMETRIC EQUATIONS

Graphing Parametric Equations

For graphing parametric equations, you'll need to change your calculator from function (Func) mode to parametric (Par) mode. Here's how: press [MODE], then use the arrow keys to select Par and press [ENTER]. (You should also check that Connected is highlighted.)

Now let's see the changes in the Y= and WINDOW screens.

- Press [Y=]. You can enter six sets of equations, XT and YT. (If you have any parametric equations stored in memory, erase them by positioning the cursor on each equation and pressing [CLEAR].)
- Press [ZOOM] [6] to set up the standard viewing window for parametric equations.
- Now press [WINDOW] to observe the WINDOW settings. Use the up [↑] and down [↓] arrow keys to scroll through the entries in this menu. Notice that the settings associated with x and y are the same as they are in function mode. However, when your calculator is in parametric mode you must also specify bounds and increments for the parameter t: Tmin, Tmax, Tstep.

Example: Graph the set of parametric equations $x(t) = 2t + 1$, $y(t) = -3t + 5$ in the standard viewing window.

- Press [Y=]. Enter the equation for x opposite X1T: press [2] [X,T,θ] [+] [1]. (Notice that, in this mode, pressing [X,T,θ] enters a T.)
- Enter the equation for y opposite Y1T: press [(-)] [3] [X,T,θ] [+] [5].
- Press [ZOOM] [6] to view the graph in the standard viewing window. You should see a line segment in the right half of your screen.
- Press [TRACE]. The cursor will mark the location when $t = 0$. The value of t and the coordinates of the point will appear at the bottom of your screen. Press the right-arrow key [→] and the cursor will jump to the location associated with $t \approx .13$ (t increases by one Tstep). Press [→] repeatedly and watch the cursor move along the line.

You will use these equations again in the next example.

Example: Graph the position of a dot as it moves along the path, $x(t) = 2t + 1$, $y(t) = -3t + 5$, at 0.5 second increments from time $t = 0$ seconds to $t = 3$ seconds.

For this example, we assume that your calculator is in parametric mode and that you have already entered this set of parametric equations in your calculator from the previous example.

- Press [MODE], highlight Dot, and press [ENTER].
- Press [WINDOW]. Adjust the parameter settings for t: Tmin = 0, Tmax = 3, and Tstep = 0.5.
- Press [TRACE]. Then press [→] and watch the dot move from one position to the next (on a screen similar to Figure 24) in 0.5 second time increments.

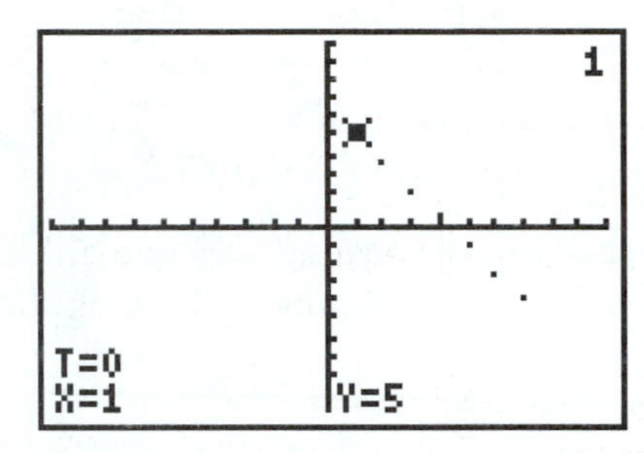

Figure 24 Graph in Dot Mode Setting

Combining Two Sets of Parametric Equations

In the lab for this section, you are asked to form a new set of parametric equations from two other sets of parametric equations.

Example: Suppose that you have two sets of parametric equations,

$$S_1:\quad x_1 = 2t + 1 \qquad\qquad S_2:\quad x_2 = t - 5$$
$$\quad\; y_1 = -3t + 5 \qquad\qquad\quad\; y_2 = 4t - 3$$

and that you want to graph the combination $(1 - t)S_1 + tS_2$ over the interval $0 \leq t \leq 1$.

Step 1: Be sure that your calculator is in parametric mode.

Step 2: Enter the two sets of parametric equations, S_1 and S_2, as X1T, Y1T and X2T, Y2T.

Step 3: Enter the x and y equations for the combination $(1 - t)S_1 + tS_2$ as follows:

$$x_3 = (1 - t)x_1 + tx_2$$

$$y_3 = (1 - t)y_1 + ty_2$$

- The cursor should be opposite X3T. Enter the equation for X3T:

On the TI-82:

Press (1 − X,T,θ) for 1 − T.

Press 2nd Y-VARS 2 for Parametric, 1 for X1T.

Press + X,T,θ 2nd Y-VARS 2 3 for +TX2T.

On the TI-83:

Press (1 − X,T,θ) for 1 − T.

Press VARS → to highlight Y-VARS and then press 2 for Parametric, 1 for X1T.

Press + X,T,θ,n VARS → 2 3 for +TX2T.

- Move the cursor opposite Y3T. Enter the equation for Y3T:

On the TI-82:

Press (1 − X,T,θ) for 1 − T.

Press 2nd Y-VARS 2 2 for Y1T.

Press + X,T,θ 2nd Y-VARS → 2 4 for +TY2T.

On the TI-83:

Press (1 − X,T,θ) for 1 − T.

Press VARS → 2 2 for Y1T.

Press + X,T,θ,n VARS → 2 4 for +TY2T.

Step 4: Graph the combination $(1 - t)S_1 + tS_2$. Here's how:

- Turn off the parametric equations for X1T, X2T, Y1T, and Y2T. Move the cursor over X1T's equals sign and press ENTER. Next, move the cursor over X2T's equals sign and press ENTER. (Notice the highlighting has also been removed from Y1T's and Y2T's equals signs.)
- Press MODE and, if necessary, highlight Connected and press ENTER.
- Press WINDOW. Set Tmin = 0, Tmax = 1, and Tstep = 0.1. Adjust the remainder of the settings for a $[-5, 2] \times [-4, 6]$ window.
- Press GRAPH. Your graph should look like Figure 25.

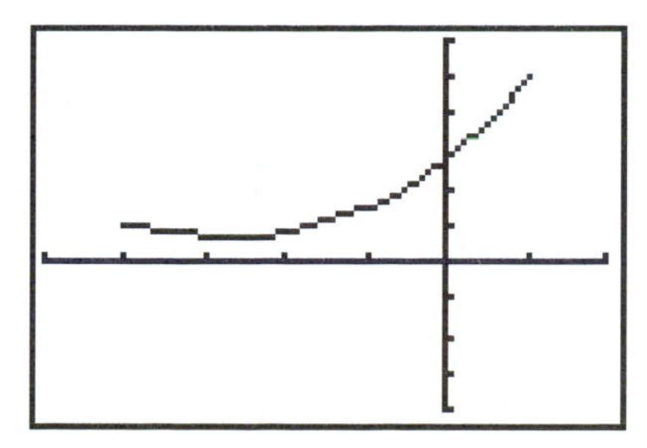

Figure 25 A Bézier Curve

Using Square Scaling in Parametric Mode

Here's how to get a window with square scaling in parametric mode.

- First, select a viewing window that shows the basic details of your graph.
- Then press [ZOOM] [5] to adjust to square scaling.

Example: Graph the set of parametric equations $x = 3\cos(t)$, $y = 3\sin(t) + 2$.

- Press [MODE] and check that Radian, Par, and Dot are highlighted. If necessary adjust these settings.
- Press [Y=] and clear any previously stored functions. Then enter the set of parametric equations.
- Press [Window]. Set Tmin to 0, Tmax to 6.3, and Tstep to 0.1. Adjust the remaining settings for a $[-6, 6] \times [-6, 6]$ window.
- Press [GRAPH]. Your graph should look egg-shaped. Next, press [ZOOM] [5] to observe the graph in a window with square scaling. Your graph should now look like a circle.

TI-85/86 GUIDE

This guide provides background on the TI-85/86 graphing calculators that will be useful for *Precalculus in Context: Projects for the Real World.* It consists of a basic tutorial followed by additional instructions relevant to each section in the laboratory manual. We do not attempt to show you everything that you can do on the TI-85/86.

BASIC TUTORIAL

As you proceed through this guide, note that specific keys that you are to press appear in a box. For example, you may be asked to press [GRAPH]. Operations corresponding to the blue or yellow lettering above the keys are indicated in brackets, [[ ]]. To access yellow upper key functions, press [2nd] followed by the key. To access blue upper key functions, press [ALPHA] and then the key.

Getting Started: On, Off, and Contrast

Turn the calculator on by pressing [ON].

You may need to adjust the contrast. Press [2nd] followed by holding down the up arrow key [↑] to darken or the down arrow key [↓] to lighten. (The four arrow keys are located directly below the [F4] and [F5] keys.)

To turn your calculator off, press [2nd] [[OFF]]. If you forget, the calculator will automatically turn off after a period of nonuse.

Calculating and Editing

The screen that displays your calculations is called the home screen. If your calculator is in a menu or displays a graph when it is turned off, it will return to the menu or graph when you turn it back on. Press [EXIT] one or more times, or press [2nd] [[QUIT]], to return to the home screen.

At times, you'll want to start with a clear home screen. To remove previous calculations, press [CLEAR]. You do *not* have to clear the home screen after each computation.

Example: Compute 3×4.

After pressing [3] [×] [4], press [ENTER]. Note that the original problem, written as $3 * 4$, remains on the left side of the screen and the answer appears to the right.

Warning! *The TI-85/86 has two minus keys,* [−] *and* [(-)], *to differentiate between the operation of subtraction (such as* $3 - 2 = 1$*) and the opposite of a number (such as* -2*). On your calculator screen, the subtraction sign appears slightly longer than the negative sign.*

Example: Compute $-2 + 5$.

To compute $-2 + 5$, press [(-)] then [2] to create the -2. Finish the computation to get the answer 3.

Note: *If you press* [−] *instead* [(-)] *in this problem, you get the wrong answer! Try it for yourself.*

Example: Compute $5 - 2$.

To compute $5 - 2$, press [5] [−] [2] [ENTER].

Example: Compute 8^2 and 1.05^7.

Press [8] followed by [x^2] [ENTER]. You can also compute the square of eight by pressing [8] [∧] [2]. Now try 1.05^7 using [∧] [7] to compute the power.

Example: Compute $\sqrt{16}$.

Press [2nd] [[√]] (same key as [x^2]) followed by [1] [6] [ENTER].

Example: Compute $\sqrt{-16}$.

Press [2nd] [[√]] followed by [(-)] (the key to the left of [ENTER]) [1] [6] [ENTER].

Many calculators would report an error message when you tried to compute $\sqrt{-16}$ because there is no real number whose square is -16. However, your TI-85/86 calculator does the computation using complex numbers. It responds: $(0, 4)$, which represents the number $0 + 4 \cdot \sqrt{-1}$ or $0 + 4i$.

Example: Compute $\sqrt[5]{32}$.

- Press [5] for the fifth root.
- Press [2nd] [[MATH]] (same key as [×]) to access the MATH menu, then press [F5] to choose MISC. The first five choices of the MATH/MISC menu are displayed on the bottom line of the menu. (Your screen should match the one in Figure 1.)

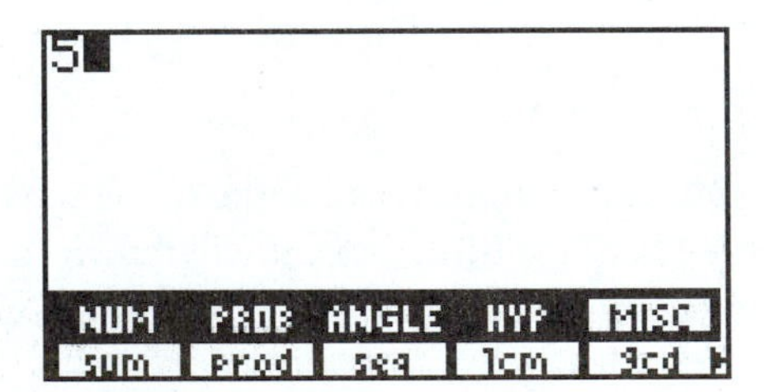

Figure 1 The MATH/MISC Menu

- Press [MORE] to reveal more options. The root operation, $\sqrt[x]{}$, will now appear above the [F4]-key. Press [F4] to select $\sqrt[x]{}$.
- Press [3] [2] [ENTER]. Did you get 2?

Correcting an Error

We tackle two situations connected with making errors. First, we look at an example that your calculator recognizes as an error. Then we provide an example that illustrates what you can do when you discover that you have punched in an error that the calculator is able to compute.

Correcting by Deleting

Example: Let's start by making a deliberate error: press [3] [+] [+] [2] [ENTER].

Your calculator will respond with an error message. Press [F1] and the cursor will GOTO the error. Erase one of the plus signs by pressing [DEL] for delete, and then [ENTER]. The correct answer to 3 + 2 will appear.

Correcting by Inserting

Example: Press [3] [+] [4] [ENTER] and suppose that you really wanted −33 + 4.

Press [2nd] [[ENTRY]] to return to the previous command. Use the left-arrow key [←] to position the cursor over the 3. Press [2nd] [[INS]]. (Notice that the cursor changes from a box to a line.) Now press [(-)] [3] to insert −3 and then [ENTER] to complete the calculation.

Resetting the Memory

Warning! *Resetting the memory erases all data and programs. All calculator settings will return to the default settings.*

Here's how to reset your calculator:

- Press [2nd] [[MEM]] [F3] to access the MEM/RESET menu. Your screen should be similar to Figure 2.

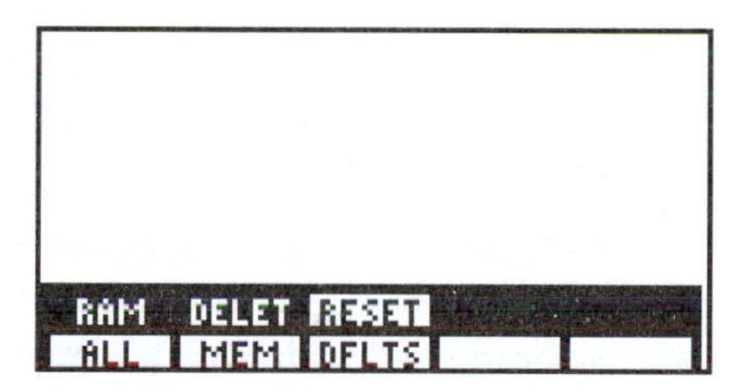

Figure 2 MEM/RESET Menu on TI-85

MEM clears the memory (programs, data, functions, graphs) but leaves the mode settings alone, DFLTS leaves the memory alone but returns the mode settings to their defaults, and ALL resets both memory and defaults.

- If you have programs or data that you do not want to erase, press [F3] to return only the mode settings to the defaults. Otherwise, press [F1] to reset the memory and return all calculator settings to the defaults. The calculator will respond: **Are you sure?** Press [F4] for Yes. The message **Mem cleared, Defaults set** should appear on your screen. If you can't read this message, darken the contrast by pressing [2nd] and then holding down the up-arrow key [↑].
- Press [CLEAR] to clear the screen.

Changing Mode Settings

Press 2nd [MODE]. The cursor should be blinking on Normal. If you have reset your calculator as shown in the previous topic, the mode settings will match the default settings shown in Figure 3.

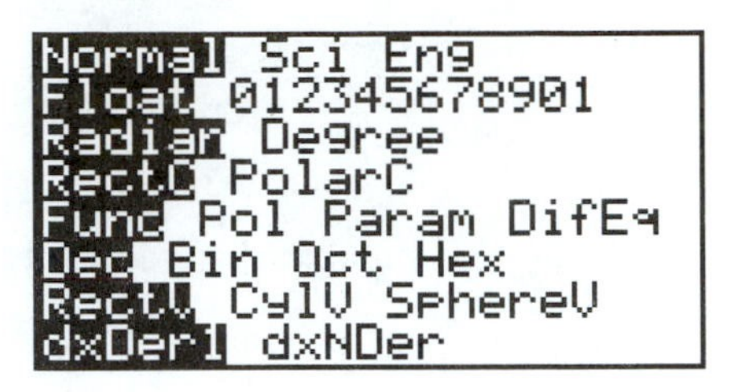

Figure 3 Mode Menu

Suppose that you want all your answers to be displayed in scientific notation. Use the right-arrow key → to highlight Sci (for scientific notation) and then press ENTER. Press EXIT to exit the mode menu.

Example: Compute 3654×1781.

The displayed answer should read: 6.507774 E 6. (E6 is shorthand notation for 10^6.)

To return the mode setting to Normal, press 2nd [MODE]. The cursor should be blinking on Normal, so press ENTER. Press EXIT to exit the mode menu.

Graphing

First, check your mode settings: press 2nd [MODE]. For now, adjust the mode settings to match the default settings in Figure 3.

Next, check to see whether any functions have been stored in your calculator; press GRAPH F1 to view the GRAPH/$y(x)$ = screen. The screen in Figure 4, for example, indicates that two functions have been stored. Here's one way to erase them.

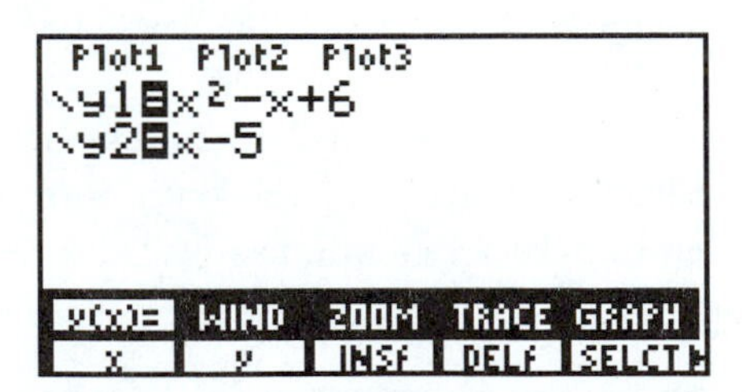

Figure 4 GRAPH/$y(x)$ = Screen

- The cursor should be blinking opposite $y1$. Press CLEAR to erase the function stored as $y1$.
- Press ↓ to move the cursor opposite $y2$, and then press CLEAR. If you need to clear more than two functions, continue the process.
- Press 2nd [QUIT] to return to the home screen.

If you did not have any functions stored in your calculator, test these instructions after you have completed the next example. As you work through this guide, unless you are told otherwise, you should erase stored functions before you begin a new example.

Example: Graph $y = x^2$ and $y = x^3$ in the standard viewing window.

- Press GRAPH F1 for the GRAPH/$y(x)$ = screen. The cursor will be blinking opposite $y1$.
- Press F1 for x followed by x^2 to enter the first function. Now press ENTER to enter the first function and move the cursor to $y2$ =.
- To enter the second function, press F1 followed by ^ 3.
- Press EXIT to return to the GRAPH menu. To graph $y = x^2$ and $y = x^3$ in the standard viewing window, press F3 for ZOOM followed by F4 for ZSTD. Press CLEAR to clear the GRAPH-menu bar from the bottom of the viewing screen.
- Press GRAPH to return the GRAPH-menu bar to the bottom of your screen. Press F2 for RANGE on the TI-85 and WIND (short for WINDOW) on the TI-86. Your screen should match one of those below.

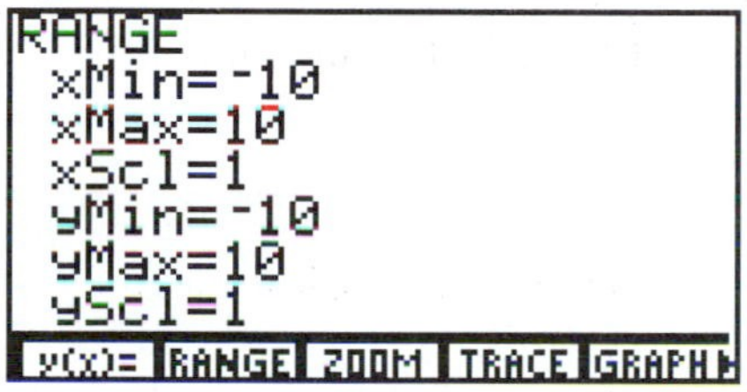

```
RANGE
 xMin=-10
 xMax=10
 xScl=1
 yMin=-10
 yMax=10
 yScl=1
y(x)= RANGE ZOOM TRACE GRAPH
```

Figure 5 Window Screen for TI-85

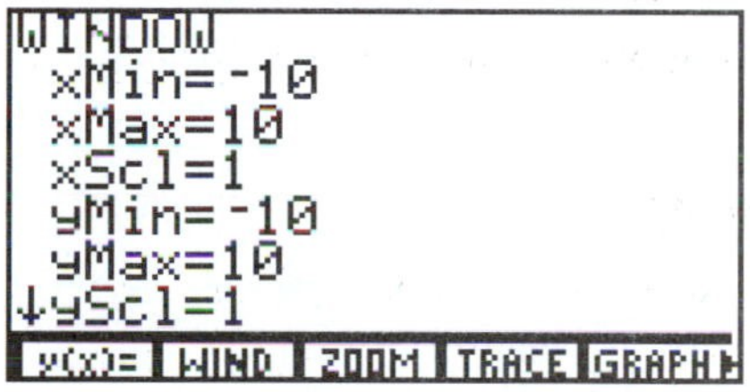

```
WINDOW
 xMin=-10
 xMax=10
 xScl=1
 yMin=-10
 yMax=10
↓yScl=1
y(x)= WIND ZOOM TRACE GRAPH
```

Figure 6 Window Screen for TI-86

In the standard viewing window, the scaling on the x- and y-axes goes from -10 to 10, with tick marks one unit apart (since Xscl and Yscl = 1). At times we may refer to this window using the notation $[-10, 10] \times [-10, 10]$.

In the next example, you will continue where the previous example left off.

Example: Graph $y = x^2$ and $y = x^3$ in the window $[-1.5, 1.5] \times [-1.0, 2.0]$. Place tick marks every 0.5 units. (In this example, the window settings should match the ones on the screen in Figure 7.)

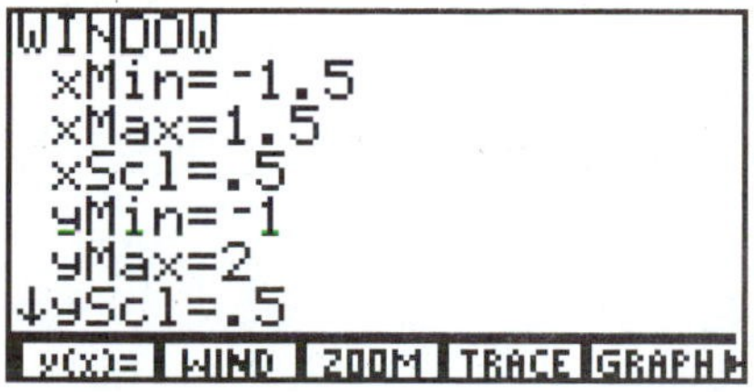

```
WINDOW
 xMin=-1.5
 xMax=1.5
 xScl=.5
 yMin=-1
 yMax=2
↓yScl=.5
y(x)= WIND ZOOM TRACE GRAPH
```

Figure 7 New Window Settings, TI-86 Screen

- You should be looking at the window screen from the previous example. Press [(-)] [1] [.] [5] [ENTER] to change the value of Xmin.
- Now, your cursor should be to the right of Xmax. Change the value of Xmax to 1.5 and press [ENTER]. Continue changing the window settings until yours match the ones in Figure 7.
- Finally, press [F5] for GRAPH.

That's it! You have completed the tutorial. Now practice and experiment on your own with the calculator until you begin to feel comfortable with these basic operations. The remainder of this guide will introduce new techniques as they are needed, section by section, for your work in *Precalculus in Context: Projects for the Real World.*

Section-by-Section Guide

SECTION 1: LINEAR FUNCTIONS

Clearing Stored Functions

Unless you are told otherwise, erase any previously stored functions each time you begin a new example. The tutorial demonstrated one method for clearing functions. Here is another.

- Press 2nd [MEM] F2 to access the MEM/DELETE menu shown in Figure 8.

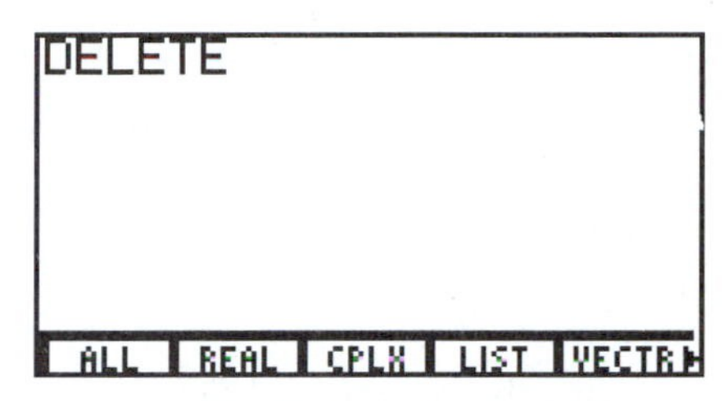

Figure 8 MEM/DELETE Menu

Note: *If you press* F1 *for ALL, you will erase* everything *that is stored in your calculator's memory, including programs, data, and functions. The MEM/DELETE menu allows you to be selective in what you choose to erase. In this case, we want to erase only the stored functions. These will be listed as equations under EQU.*

- Press MORE to view other items in this menu. EQU should appear above the F3 key. Press F3 to select EQU. A list of all stored functions will appear.
- Each time that you press ENTER, you will delete the function to the right of the triangle marker. Press ENTER repeatedly until all functions have been cleared or use the arrow keys to selectively mark functions that you wish to clear.

Plotting Points

You can use your calculator to plot the Fahrenheit-Celsius data given in the preparation for Lab 1A. Then graph your guess for the formula that relates degrees Fahrenheit to degrees Celsius. This will allow you to check how closely the function specified by your formula follows the pattern of the data.

Example: Plot the data in the following table and then overlay the graph of $y = 18x + 85$.

Sample Data	
x	y
−2	40
−1	60
1	100
3	140

Step 1: Clear any stored functions. (See *Clearing Stored Functions.*)

Step 2: The TI-85/86 has built-in list names for the x-variable list (xStat) and the y-variable list (yStat). First, we check to see if any data have been stored in these lists and, if so, clear the lists as follows.

On the TI-85:

- Press STAT F2 for EDIT. Press ENTER ENTER to get to the xStat and yStat lists. If no data have been previously stored, the cursor will be blinking to the left of a blank "$x1 =$" followed by $y1 = 1$. In this case proceed to Step 3. Otherwise, complete Step 2.
- If data are stored in these lists, erase the data by pressing F5 for CLRxy.

On the TI-86:

- Press 2nd [STAT] (same key as +) followed by F2 for EDIT. You should see three lists: xStat, yStat, and fStat. If no numbers appear in the xStat and yStat lists, you can proceed directly to Step 3. Otherwise, complete Step 2.
- To erase the data in the xStat list, use the arrow keys to highlight xStat. Then press CLEAR followed by ENTER. Now do the same for the yStat list. (If there is data in the fStat list, a frequency list, you may erase it also.)

Step 3: Enter the data from the table.

After Step 2, you should be in the xStat and yStat lists. Your screen should resemble either Figure 9 or Figure 10.

Figure 9 Stat Lists on TI-85

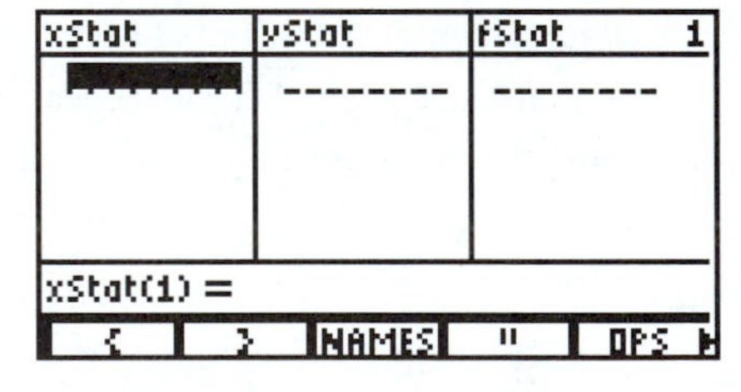

Figure 10 Stat Lists on TI-86

On the TI-85:

- Enter the data *one* x-y pair at a time. (The cursor will be blinking opposite $x1$.) Press (-) 2 ENTER 4 0 ENTER for the first ordered pair.
- Then press (-) 1 ENTER 6 0 ENTER for the second.
- Press 1 ENTER 1 0 0 ENTER for the third; and press 3 ENTER 1 4 0 ENTER for the last ordered pair.

On the TI-86:

- If necessary, use your arrow keys to highlight the cell directly beneath *x*Stat (as shown in Figure 10). Enter the *x*-data: press [(-)] [2] [ENTER] for the first entry. Then enter the remaining three values for *x*.
- Press [→] to highlight the cell directly beneath *y*Stat. Then enter the *y*-data.

Step 4: Plot the sample data.

On the TI-85:

- First, you'll need to adjust the window settings: press [GRAPH] [F2] to access the GRAPH/RANGE screen. Choose a value for *x*Min that is smaller than the *x*-coordinates in the table and a value for *x*Max that is larger than the *x*-coordinates. Similarly, select appropriate settings for *y*Min and *y*Max. Decide on the spacing of the tick marks and set *x*Scl and *y*Scl. (What would be the disadvantage of setting *y*Scl=1? How many tick marks would appear between 40 and 140?)
- Next, you'll plot the data in the window that you have selected. Press [STAT] [F3] to select DRAW and then [F2] for SCAT (scatter plot). Press [CLEAR] to clear the menu bar from the scatter plot. (If you have chosen appropriate settings for *x*Min, *x*Max, *y*Min, and *y*Max, you should see a plot of the four data points on your screen.)

On the TI-86:

- Press [2nd] [STAT] and then [F3] to select PLOT.
- Press [F1] to choose PLOT 1. To turn the plot on, press [ENTER]. Your screen should match the one in Figure 11. (If the Type does not match the one below, press [↓] and then [F1] for SCAT. You can change the lists and the symbol for the mark in a similar fashion.)

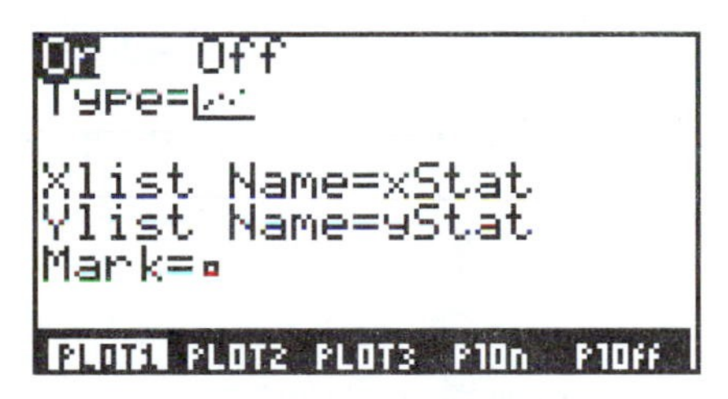

Figure 11 Screen for Plot 1

- Next, adjust the window settings: press [GRAPH] [F2] for WIND. Choose a value for *x*Min that is smaller than the *x*-coordinates in the table and a value for *x*Max that is larger than the *x*-coordinates. Similarly, select appropriate settings for *y*Min and *y*Max. Decide on the spacing of the tick marks and set *x*Scl and *y*Scl. (What would be the disadvantage of setting *y*Scl=1? How many tick marks would appear between 40 and 140?)
- Press [F5] [CLEAR] to view the scatter plot in the window you have selected.

Step 5: Add the graph of $y = 18x + 85$ to the plot of the sample data.

- Press GRAPH F1 to enter the GRAPH/$y(x)$ = screen and enter $18x + 85$ opposite $y1=$. (We assume that you have erased all other functions.) Press EXIT F5 CLEAR. On the TI-86, you will see the graph of the line and your scatter plot in the same screen. If you are using a TI-86, proceed to Step 6. However, on the TI-85, the scatter plot has disappeared from the screen. If you are using a TI-85, complete the remainder of this step.
- To overlay the scatter plot, press STAT F3 F2 CLEAR. The plotted points should lie close to the line.

Warning! *If you are using a TI-86 and you fail to turn off PLOT 1, this plot will appear superimposed on your next graph. Worse, if you subsequently erase the data in xStat and yStat without turning off this plot, you will get error messages each time you press* GRAPH *because your calculator will still be trying to create STAT PLOT 1.*

Step 6: If you are using a TI-86, press 2nd STAT F3 F5 ENTER to turn off your plots. If you are using a TI-85, the scatter plot will disappear after you have changed your function or window settings.

Adjusting the Viewing Window for Square Scaling

The viewing screen on your calculator is a rectangle. Therefore, if you use the standard window, the tick marks on the y-axis will be closer together than those on the x-axis. For square scaling we want the distance between 0 and 1 on the x-axis to be the same as the distance between 0 and 1 on the y-axis.

Example: Graph the line $y = x$ in the standard viewing window and then switch to square scaling.

Step 1: Graph $y = x$ in the standard viewing window.

- Erase any previously stored functions. (See *Clearing Stored Functions*, page 259.) Then enter $y = x$ as $y1$ and press EXIT to return to the GRAPH menu.
- Press F3 to access the GRAPH/ZOOM menu, and F4 for ZSTD.
- Press CLEAR to remove the menu bar from the bottom of the viewing screen.

Observe the spacing between the tick marks on the x- and y-axes. Notice that the tick marks on the y-axis are closer together than the tick marks on the x-axis and that the graph does not make a 45° angle with the x-axis.

Step 2: Change to a square viewing window.

- Press GRAPH F3 to access the GRAPH/ZOOM menu, and then press MORE F2 for ZSQR. When the graph appears on your screen, observe the equal distance between tick marks on the two axes. Notice that the graph makes a 45° angle with the x-axis.
- Press EXIT F2 and note the changes in the viewing window settings.

SECTION 2: QUADRATIC FUNCTIONS

Graphing a Quadratic Function

When you graph a quadratic function, it is important to experiment with various viewing windows to ensure that you have captured all the important features of the graph on your screen.

Example: Graph $y = 2x^2 - 2x + 12$ using the three different viewing windows specified in the directions below.

Start by entering the function:

- Press [GRAPH] [F1] and enter $2x^2 - 2x + 12$. (Remember to use [−] when you subtract $2x$.)
- Press [EXIT] to return to the [GRAPH] menu. Then press [F2].

Window 1: Adjust the settings to match those shown below in Figure 12 and then press [F5]. Your graph should be the familiar ∪-shape of a parabola.

Figure 12 Settings for Window 1

Next, we view the graph of this quadratic function in two other windows. In each case, the viewing window selected fails to show key features of the parabola.

Window 2: Change the xMin setting to 2 as follows.

- In the [GRAPH] menu, press [F2].
- Press to change xMin to 2 and then press [F5] to graph the function. Notice that in Window 2 the graph of $y = 2x^2 - 2x + 12$ looks more like a line than a parabola.

Window 3: Change to the standard viewing window as follows.

- While still in the graph menu, press [F3] to access the GRAPH/ZOOM menu and then press [F4] for ZSTD.
- Press [CLEAR] to remove the menu bar from the graphing screen. What do you see? In general, if you ask the calculator to graph a function, and just see empty axes, you probably are looking at a part of the plane that contains none of the graph.

Calculating Outputs of a Function

Example: Find the output of $f(x) = x^2 - x + 6$ when x has value 4.

- Press GRAPH F1 and enter $x^2 - x + 6$ as $y1$.
- Press EXIT MORE MORE F1 for EVAL.
- Enter the x value: press 4 ENTER. The corresponding y-value of 18 will appear at the bottom of your screen.

Now, use the method outlined above to check that $f(-2) = 12$. (First, you'll need to press EXIT to return to the menu containing EVAL.)

Using TRACE and ZBox to Approximate the Coordinates of a Point on the Graph

Example: Approximate the vertex (turning point) of the parabola $y = 2x^2 + 3x - 2$.

Plan of action: First, we'll graph $y = 2x^2 + 3x - 2$ in the standard viewing window. Then we'll use ZBox to get a window that magnifies the section of the graph containing the vertex and x-intercepts. Finally, we'll use TRACE to estimate the vertex and x-intercepts.

Step 1: Enter and graph $y = 2x^2 + 3x - 2$ in the standard viewing window.

- Press GRAPH F1 to access the GRAPH/$y(x)$ = screen and enter the function $y = 2x^2 + 3x - 2$ as $y1$.
- Press EXIT F3 F4 to graph the quadratic in the standard viewing window. Your graph should resemble the one in Figure 13 (without the box).

Step 2: Magnify the section of graph containing the vertex and x-intercepts by drawing a box that will become the new viewing window.

- You should still be in the GRAPH menu. Press F3 for ZOOM and then press F1 for Box. There should be a blinking cursor (a plus sign with a blinking pixel in the middle) at the center of your screen. (You may have trouble seeing the cursor until you move it off the origin.)
- Next, we are going to draw a box like the one in Figure 13. Start by pressing the left ← and up ↑ arrow keys to move the cursor to the location where we want the upper left corner of the box. Then press ENTER.
- Now use the right → and down ↓ arrow keys to draw a box similar to the one in Figure 13. Press ENTER to replace the window settings with those that correspond to the box.

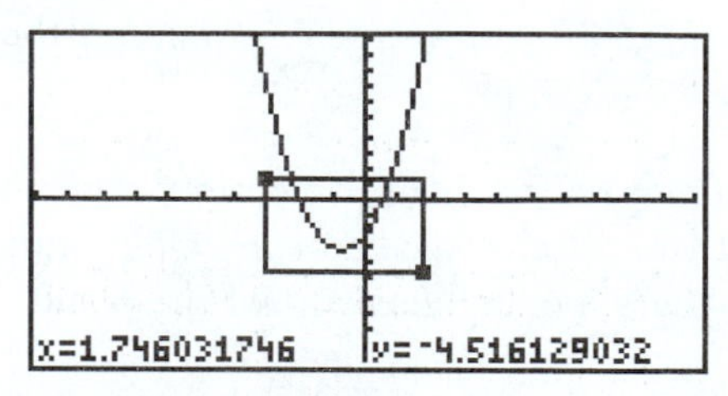

Figure 13 Graph of $y = 2x^2 + 3x - 2$

Step 3: Use TRACE to find the approximate coordinates of the x-intercepts and vertex.

- Press EXIT to return to the GRAPH menu and then F4 for TRACE. Use the right → and left ← arrow keys to position the cursor at the vertex of the parabola. Then read the approximate coordinates of the vertex at the bottom of your screen.
- Continue to use the right → and left ← arrow keys to position the cursor on each of the x-intercepts. Read the approximate coordinates for each of the x-intercepts.

Your answers should be similar to those in the screens in Figure 14.

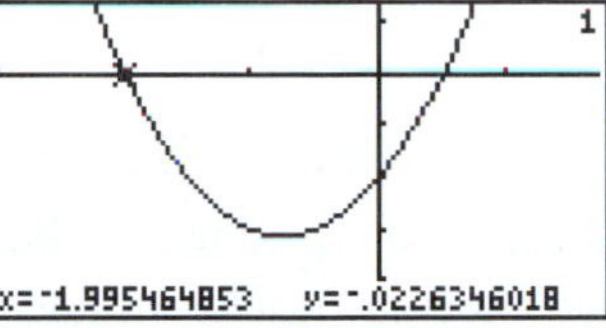

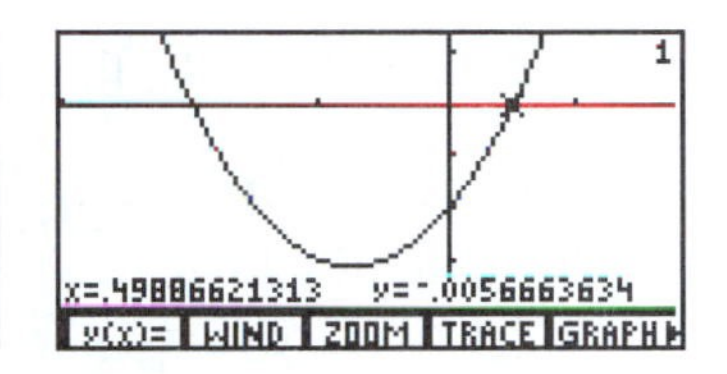

Figure 14 Approximate Coordinates of the Vertex and x-intercepts

SECTION 3: SHIFTING AND REFLECTING GRAPHS

In Lab 3, *Graph Trek Explorations*, you will be investigating the effect that certain algebraic modifications, such as adding a constant to the input variable, have on the graph of a function. You'll want to experiment using several different functions. We've provided some functions and algebraic modifications that you might want to consider.

Using Parentheses

Warning! *When you want to apply a function to an expression, you must enclose the entire expression in parentheses.*

Example: Graph $y = \sqrt{x+2}$ in the window $[-5, 5] \times [-5, 5]$.

- Enter $y = \sqrt{x+2}$ as $y1$: press GRAPH F1 and after clearing any previously stored functions position the cursor to the left of $y1$. Then press 2nd [√] (F1 + 2).
- Press EXIT F2 and adjust the settings for a $[-5, 5] \times [-5, 5]$ window. Then press F5 to view the graph. Your graph should resemble the one in Figure 15.

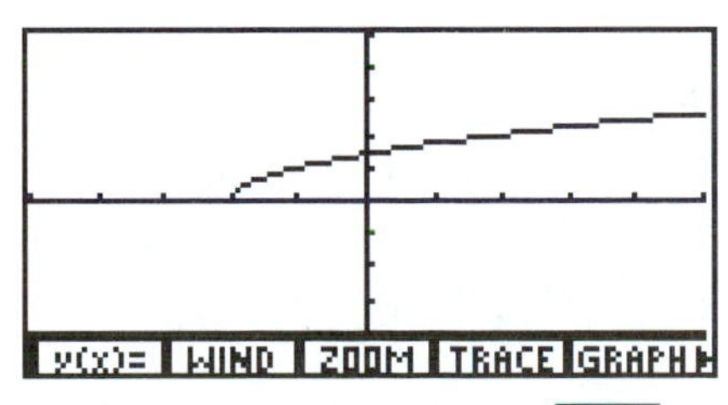

Figure 15 Graph of $y = \sqrt{x+2}$

Graphing Trigonometric Functions in the Trig Viewing Window

Locate the SIN, COS, and TAN-keys on your calculator. These keys will allow you to study the graphs of the sine, cosine, and tangent functions before the functions are formally introduced. Before entering any of these functions, press 2nd [MODE]. If Radian is not highlighted, move the cursor to Radian and press ENTER.

Example: Examine the graphs of $y = \sin(x)$ and $y = \sin(x + 2)$ in the trig viewing window.

- Press GRAPH F1 and enter the function $\sin(x)$ as $y1$: press SIN (F1) ENTER.
- Next, enter the function $\sin(x + 2)$ as $y2$: press SIN (F1 + 2).
- After pressing EXIT to return to the GRAPH menu, press F3 MORE F3 for ZTRIG. Press CLEAR to view the graph without the menu bar. You should see two wavy curves. (If you don't, go back and check that you are in Radian mode.)
- Finally, press EXIT then F2 and observe the settings for the trigonometric window.

Using the Absolute Value Function

Example: Graph $y = |x|$ in the standard viewing window.

- Press GRAPH F1 for the GRAPH/$y(x)$ screen. Enter $y = |x|$ as follows:

On the TI-85:

Press 2nd [CATALOG]. (The triangle marker should already be pointing to abs.) Press ENTER to select abs (short for the absolute value) and then (F1).

On the TI-86:

Press 2nd [CATLG-VARS] followed by F1 to select the CATALOG. (The triangle marker should already be pointing to abs.) Press ENTER to select abs (short for the absolute value) and then (F1).

- Press EXIT F3 F4 to view the V-shaped graph of the absolute value function.

Graphing a Family of Functions

Using your calculator's list capabilities, you can substitute each value in a given list for a constant in an algebraic formula. This feature allows you to graph an entire family of functions quickly. On the TI-85/86, you specify a list by enclosing the members of the list in brackets: { }.

Example: Graph the family of quadratic functions $y = (x + 1)^2$, $y = (x + 2)^2$, and $y = (x + 3)^2$ in the window $[-5, 5] \times [-5, 5]$.

- Press GRAPH F2 and adjust the settings for a $[-5, 5] \times [-5, 5]$ window.
- Press F1 to access the GRAPH/$y(x)$= screen. Then enter the three functions by specifying the constants, 1, 2, and 3, in a list as follows. Press (F1 + 2nd [LIST] (same key as −) and then F1 for {. Next, press 1 , 2 , 3 and then F2 for }. Press) x^2.
- Press EXIT twice to return to the GRAPH menu, then press F5. Watch as the three functions are graphed one after the other.

SECTION 4: EQUATIONS AND INEQUALITIES

Finding the Points of Intersection of Graphs

Example: Find the points where the graphs of $f(x) = -2x^2 + 2x + 12$ and $g(x) = -4x + 15$ intersect.

You could approximate the solutions to this problem using the techniques outlined in the example on page 264. However, the TI-85/86's intersect (ISECT) command will provide, when possible, exact solutions and, if not possible, very good approximations.

Step 1: Graph both functions in a viewing window that gives a clear view of the points of intersection.

- Enter the two functions into your calculator.
- Adjust the window settings so that you can see both points of intersection. (*Hint:* you might start with the standard viewing window, ZSTD in the GRAPH/ZOOM menu, and then adjust the window settings after looking at the graph.)

Step 2: Approximate the coordinates of one of the points of intersection.

- From the GRAPH menu press MORE and then F1 to access the GRAPH/MATH menu. Press MORE again. Next, press the F-key that corresponds to ISECT (F5 on the TI-85 and F3 on the TI-86). Your screen should resemble Figure 16.

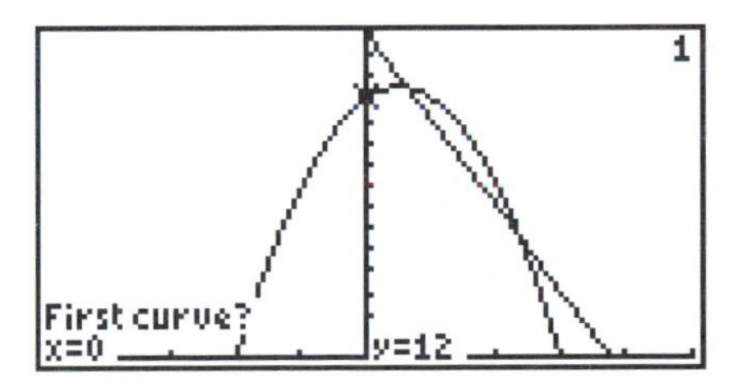

Figure 16 The Intersection of Two Graphs

- Try pressing the up ↑ or down ↓ arrow keys. The cursor will jump back and forth from the line to the parabola. Position the cursor on the line (We'll designate the line to be the first curve), and then press ENTER. The cursor will automatically jump to the other curve. If you are using a TI-86, press ENTER once more to select the parabola for the second curve.

- Now you must provide your calculator with a guess for the point of intersection. Press [→] or [←] to move the cursor close to one of the points of intersection and then press [ENTER]. Read off the (approximate) coordinates for this point of intersection.

Step 3: Find the coordinates of the second point of intersection by repeating the process outlined in Step 2.

If you have done everything correctly, you will find that the two graphs intersect at approximately (0.634, 12.464) and (2.366, 5.536).

Graphing Piecewise Defined Functions

Example: Graph the piecewise defined function $f(x) = \begin{cases} x - 4 \text{ if } x > 4 \\ -x + 4 \text{ if } x \leq 4 \end{cases}$

The graph of $f(x)$ consists of two half-lines pieced together. You'll want the graph of $y = x - 4$ when x-values are greater than 4 and $y = -x + 4$ when x-values are less than or equal to 4.

Step 1: Enter the functions that you want to piece together as follows:

- From the GRAPH/$y(x)$= screen, enter $y = x - 4$ as $y1$ and $y = -x + 4$ as $y2$.
- Press [EXIT] [F3] [F4] to graph the two functions. Your graph should look like a cross: ×.

Step 2: Form the function $f(x)$. This is where you'll piece together the graphs of $y1$ and $y2$.

- Return to the GRAPH/$y(x)$= screen. Press [ENTER] twice to position the cursor opposite $y3$.

Here's how to enter $f(x)$ as $y3$:

- To insert $y1$: press [F2] [1].
- Press [×]. Next, enter the condition that governs when to use $y1$: press [(] [F1] [2nd] [[TEST]] (same key as [2]) followed by [F3] [4] [)]. Then press [EXIT] to return to the GRAPH/$y(x)$= menu.
- Press [+].
- To insert $y2$: press [F2] [2].
- Press [×]. Then enter the condition that governs when to use $y2$: press [(] [F1] [2nd] [[TEST]] [F4] [4] [)]. When you have completed this step your screen should look similar to Figure 17. (On the TI-86 you will see three dots to the right of $y3$. If you press the right-arrow key, you will see the entire function.)

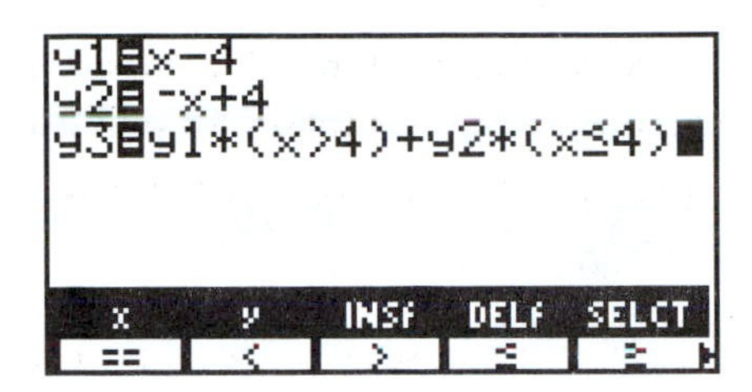

Figure 17 GRAPH/$y(x)$ = Screen on TI-85

Here's how your calculator interprets the information you've just entered as a piecewise defined function. The calculator assigns the expression ($x > 4$) the value when the inequality is true (in other words, when the input variable, x, is greater than 4). In this case, the inequality $x \leq 4$ is false, so the calculator sets the expression $x \leq 4$ equal to 0. Thus, for $x > 4$, the function $y3$ is equivalent to:

$$y3 = (x - 4)(1) + (-x + 4)(0) = x - 4.$$

And when $x \leq 4$, the function $y3$ is equivalent to:

$$y3 = (x - 4)(0) + (-x + 4)(1) = -x + 4.$$

Step 3: Graph $f(x)$.

- Press EXIT to return to the GRAPH/$y(x)$= menu.
- Turn off $y1$ and $y2$ as follows. Position the cursor opposite $y1$=. Press F5 for SELCT (select) to remove the highlighting from $y1$'s equals sign. Then position the cursor opposite $y2$= and press F5.
- Press EXIT F5. The graph of $f(x)$ should look V-shaped.

Fitting a Line to Data

If your data, when plotted, lie exactly on a line, you can use algebra to determine the equation of the line. However, real data seldom fall precisely on a line. Instead, the plotted data may exhibit a roughly linear pattern. The least squares line (also called the regression line) is a line that statisticians frequently use when describing a linear trend in data.

Example: Use the least squares line to describe the linear pattern in the data below.

x	y
−3.0	−6.3
−2.0	−2.8
1.2	2.0
2.0	4.1
3.1	5.0
4.2	7.2

Step 1: Erase any stored functions and stored data in the xStat and yStat lists. Then enter the data from the table above into the xStat and yStat lists. (Refer to *Plotting Points* on pages 259–262.)

Step 2: Next, we'll find the equation for the least squares line. Statisticians specify the form of this line as $y = a + bx$. Here b represents the slope of the line, and a is the y-intercept.

On the TI-85:

- After you have entered your data, Press [EXIT] [F1] to select CALC.
- Press [ENTER] twice to access the STAT/CALC menu. Then press [F2] for LINR.

On the TI-86:

- After you have finished entering your data, press [2nd] [[EXIT]] to return to the home screen. Then press [2nd] [[STAT]] [F1] to select CALC.
- Press [F3] for LinR. Then press [2nd] [[LIST]] [F3] to access the names of your lists. Finally, press [F2] [,] [F3] [ENTER].

If you have entered your data correctly, your output will include $a = -0.09952028$ and $b = 1.78129485$.

Step 3: Graph the least squares line (approximately $y = -0.0995 + 1.7813x$) and the data in the same viewing window.

- Press [GRAPH] [F1]. The cursor should be opposite $y1$.
- Insert the equation for the least squares line. Here's how:

On the TI-85: Press [2nd] [[VARS]] [MORE] [MORE] [F3] to select STAT. Use the down-arrow key [↓] to move the marker opposite RegEq, and then press [ENTER].

On the TI-86: Press [2nd] [[STAT]] [F5] for VARS. Now press [MORE] [MORE] [F2] to select RegEq.

You should see $y1$ = RegEQ at the top of your screen.

- Now graph the least squares line and the scatter plot of data as follows:

On the TI-85:

Press [EXIT] once to access the GRAPH menu. Press [F2] and adjust the window settings so that all data points, when plotted, will appear on your screen. Now, press [STAT] [F3] for DRAW and a graph of the least squares line will appear on your screen. Finally, press [F2] to add the scatter plot to the display.

On the TI-86:

Press 2nd [QUIT] to return to the home screen. Then press 2nd [STAT] F3 for PLOT. Press F1 and then ENTER to turn PLOT 1 on. Now, press GRAPH F3 MORE F5 to graph both the line and scatter plot in the data viewing window.

- Press CLEAR to remove the menu bar from the bottom your screen. Your graphs should be similar to the one in Figure 18. (The scatter plot will be dots on the TI-85 and hence will be more difficult to see.)

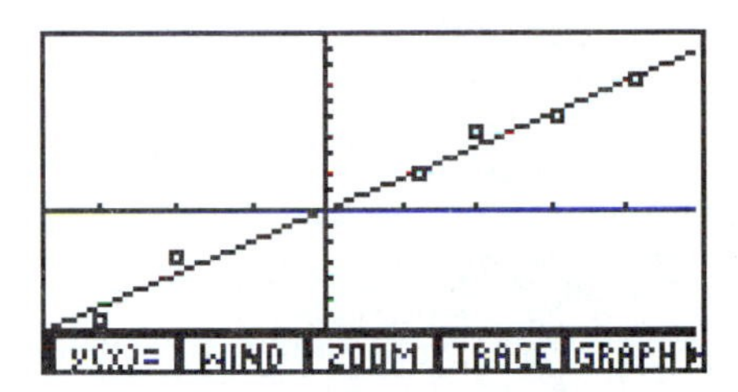

Figure 18 Scatter Plot and Regression Line

- On the TI-85, remove the scatter plot from your screen by pressing STAT F3 F5 for CLDRW (Clear Draw). On the TI-86, press 2nd [STAT] F3 F5 ENTER to turn off PLOT 1.

SECTION 5: POLYNOMIAL FUNCTIONS

Determining the Roots of a Function

Determining the roots (zeros, or x-intercepts) of a function using factoring often requires considerable skill; and most polynomials can't be factored easily, if at all. However, your calculator can take much of the drudgery out of finding roots.

Example: Find the roots of the polynomial function $g(x) = \frac{1}{2}x^4 + 2x^3 + x^2 - 3x - 5$.

Step 1: Graph $g(x)$ in the standard viewing window as follows.

- Enter $g(x)$ as $y1$. (Note: The entire function will not fit on your screen. Instead of the first term of the polynomial, you will see ellipsis marks (. . .). If you want to view the first term, move the cursor to the front of the polynomial by pressing 2nd ←.)
- Press EXIT F3 F4 to view the graph in the standard viewing window.

Step 2: Let's approximate the negative root first. You could zoom in on this intercept and then use TRACE to estimate the x-coordinate. However, here is another way to determine the root.

- From the GRAPH menu, press MORE F1 to access the GRAPH/MATH menu.

On the TI-85:

To approximate the negative root, you'll have to specify a narrow interval about the negative x-intercept. Press F1 for LOWER. Using the left-arrow key ←, move the cursor slightly to the left of the negative x-intercept and then press ENTER to mark with a black triangle the lower bound of the interval.

Next, press F2 for UPPER. Use the right-arrow key → to position the cursor slightly to the right of the negative root (*but still to the left of the positive root*) and press ENTER, this time marking the upper bound of the interval.

Now, press F3, for ROOT.

On the TI-86:

Press F1 for ROOT.

To approximate the negative root, you'll have to specify a narrow interval about the negative x-intercept. Press ← to move the cursor to a spot on the curve to the left of the negative x-intercept and press ENTER. Next, press → to move the cursor slightly to the right of the negative root (*but still to the left of the positive root*) and press ENTER.

- You'll have to provide the calculator with a guess for the root: use ← and/or → to move the cursor as close to the negative root as possible and then press ENTER.
- Read off the approximate value of the root (or zero) from the bottom of your viewing screen. The x-value should be close to -3.05 and the corresponding y-value very close to zero. (Note that $-5E-13$ is -0.0000000000005.)

Step 3: To approximate the positive root, press GRAPH and then adapt the instructions in Step 2. (On the TI-85, the marker triangles will move when you select new upper and lower bounds. They will disappear from your screen if you alter any of the window settings.) After adapting the instructions, you should get approximately 1.39 for this root.

Finding Local Maxima or Minima of Functions

In Section 2, you found the vertex of a parabola using a procedure requiring a combination of ZOOM/BOX and TRACE. (See page 264 for details.) The coordinates of any function's turning points could be estimated using this method. One drawback to this procedure is that you must frequently apply it several times in succession before you can obtain the desired accuracy. Here is another method for approximating local maxima and/or minima of a function.

Example: Let's estimate the local maximum and local minimum of the cubic function $f(x) = x^3 - 4x^2 + 2x - 4$.

Step 1: Graph $f(x)$ using a viewing window that gives you a clear view of the two turning points (one peak and one valley) of the graph.

Step 2: Approximate the coordinates of the turning point associated with the local maximum (the y-coordinate of the peak) as follows.

- In the GRAPH menu, press MORE F1 for the GRAPH/MATH menu.

On the TI-85:

Use LOWER and UPPER to specify an x-interval that contains the x-coordinate associated with the local maximum of the function. (Refer to Step 2 of *Determining the Roots of a Function* for help in setting up this interval.) While still in the GRAPH/MATH menu, press MORE F2 for FMAX.

On the TI-86:

Press F4 for FMAX. Specify an x-interval that contains the x-coordinate associated with the local maximum of the function. (Refer to Step 2 of *Determining the Roots of a Function* for help in setting up this interval.)

- Your calculator needs a guess for the turning point. So, you'll need to use ← and/or → to position the cursor on the turning point where the graph peaks and then press ENTER. Your screen should look similar to Figure 19, which shows a local maximum value of approximately -3.73.

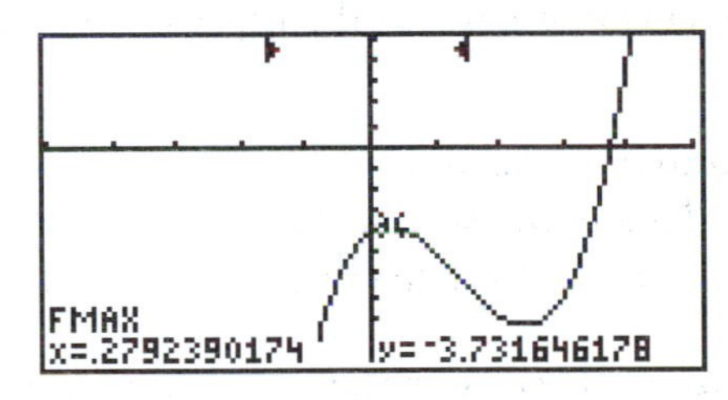

Figure 19 Approximation of Local Maximum

Step 3: Approximate the coordinates of the turning point associated with the local minimum (the y-coordinate of the valley on the graph).

- Press GRAPH MORE F1 to access the GRAPH/MATH menu.
- Adapt the instructions for Step 2 using FMIN. You should get a value for y that is close to -8.42.

SECTION 6: RATIONAL FUNCTIONS

Graphing a Rational Function

Warning! *If the numerator or the denominator of a rational function consists of more than one term, you must enclose it in parentheses when you enter it into your calculator.*

Example: Graph $r(x) = \dfrac{x^2 - 1}{x - 3}$.

- Enter $r(x)$ as $y1$: press [Graph] [F1] and then [(] [F1] [x^2] [−] [1] [)] [÷] [(] [F1] [−] [3] [)].
- Press [EXIT] [F3] [F4] to view the graph in the standard window. Press [CLEAR]. Your graph should resemble the one in Figure 20.

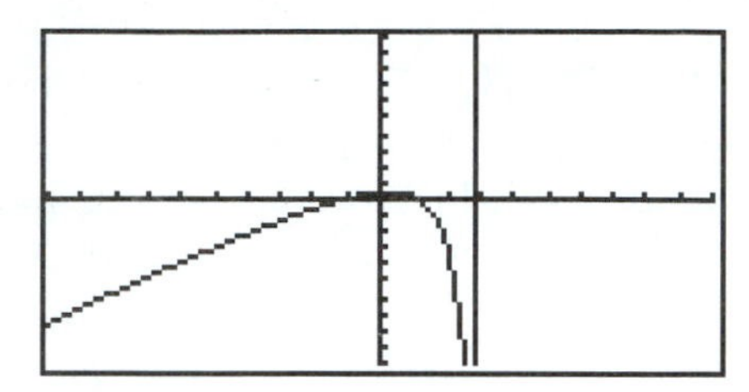

Figure 20 Graph of $r(x)$ in Standard Viewing Window

The vertical line in the graph above indicates that this function has a vertical asymptote at $x = 3$. Remember, this line is not part of the graph of the function. Furthermore, because the domain of this function is all real numbers except $x = 3$, there is a branch of this graph that lies to the right of the line $x = 3$. In order to observe this branch, you would have to adjust the Window setting for yMax.

Do *not* clear this function from your calculator's memory until after you have completed the next topic, *Zooming Out.*

Zooming Out

Example: Graph the function $f(x) = \dfrac{x^2 - 1}{x - 3}$ in the standard viewing window and then zoom out by a factor of 4 several times.

For this example, observe what happens to the appearance of the graph of $r(x)$ as we "back away" by increasing the width and height of the viewing window. You should have the graph of $r(x)$ (from the previous example) in your screen.

- Press [GRAPH] [F2] to access the window settings. Set xScl and yScl equal to 0. (This turns off the tick marks that appear on the axes. If you skip this step, the axis will get crowded with tick marks when you zoom out.)
- Check the settings for the zoom factor: press [F3] to access the GRAPH/ZOOM menu, and then press [MORE] [MORE] and the F-key corresponding to ZFACT ([F1] on the TI-85 and [F2] on the TI-86). If necessary, adjust the factor settings so that xFact = 4 and yFact = 4.
- Press [F3] to re-enter the GRAPH/ZOOM menu, then press [F3] a second time for ZOUT. (A blinking pixel, part of a free moving cursor, should appear in the center of your screen. The zooming will be centered around this location. If you wish to change the focal point of the zooming, use the arrow keys to position this cursor at a different center.)

- Press [ENTER] to view the graph over wider x- and y-intervals. Press [ENTER] again to zoom out a second time. (The graph should look like a line except, perhaps, for a small blip slightly to the right of the origin.)
- Now, press [GRAPH] [F2] to observe the effect on the window settings of twice zooming out by a factor of 4. (The default setting for ZOUT widens both the x- and y-intervals by a factor of 4 each time that it is applied. In the previous example, you zoomed out twice. Therefore, the x- and y-intervals are 16 times wider than they were before you zoomed out.)

In the next example we'll change the default zoom settings in order to observe a graph of a function that begins to act like its horizontal asymptote.

Example: Graph $q(x) = \dfrac{5x^2 + 20x - 105}{2x^2 + 2x - 60}$ in the standard viewing window.

If you have entered the function correctly, your graph should look like Figure 21.

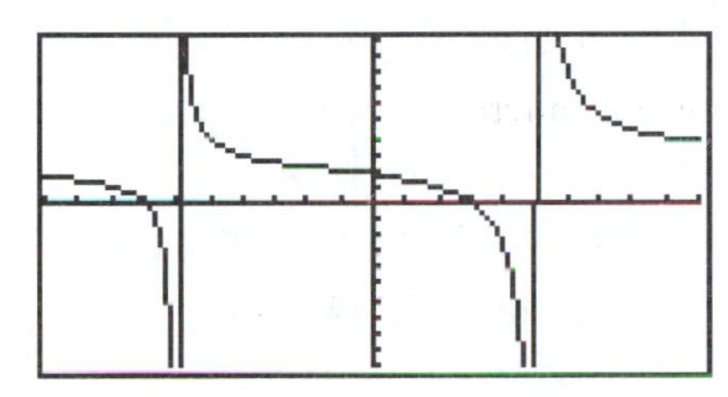

Figure 21 Graph of $q(x)$

Based on this graph you may suspect that the function $q(x)$ has a horizontal asymptote but this is not at all obvious. Let's observe the function's graph over increasingly wide x-intervals to see if it begins to behave like a horizontal line. Instructions on how to zoom out in just the horizontal direction follow.

- From the GRAPH menu, press [F2] and set xScl and yScl to 0.
- Press [F3] to access the GRAPH/ZOOM menu. Press [MORE] twice, followed by the F-key for ZOOMX ([F2] on the TI-85 and [F3] on the TI-86). Then press [ENTER]. (ZOOMX widens the x-interval by a factor of 4 and leaves the y-interval unchanged.)
- Press [ENTER] several more times to continue widening the x-interval. Your graph should begin to resemble its horizontal asymptote $y = 2.5$.

SECTION 7: EXPONENTIAL GROWTH AND DECAY

Graphing Exponential Functions

Because exponential functions increase or decrease very quickly in certain regions of their domain, you may have to experiment in order to find a viewing window that captures the function's important graphical features.

Example: Graph $y = 4^x$ in the standard viewing window. Then change the window to $[-2, 2] \times [0, 10]$.

- Press GRAPH F1. Then enter $y = 4^x$ by pressing 4 ^ F1 or 4 ^ x-VAR.
- Press EXIT F3 F4 to graph this function in the standard viewing window.

Notice that, to the left of the y-axis, the graph appears to merge with the line $y = 0$ and, to the right of the y-axis, the graph becomes so steep that it appears vertical.

- Press F2 to access the window settings. Now adjust the settings for a $[-2, 2] \times [0, 10]$ window so that the graph will fill more of the screen. Your graph should resemble the one in Figure 22.

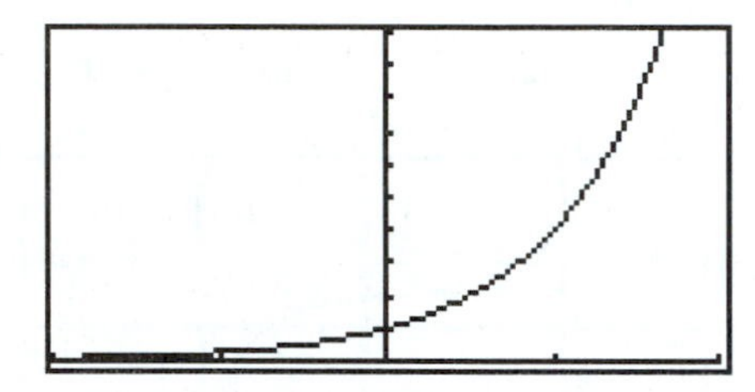

Figure 22 Graph of $y = 4^x$

There are two bases for exponential functions, 10 and e, that are so common they have their own function keys on the calculator $[10^x]$ and $[e^x]$.

Example: Graph $f(x) = e^x$ in the window $[-2, 3] \times [-1, 12]$.

- Press GRAPH F1 and 2nd $[e^x]$ (same key as LN) (F1).
- Press EXIT F2 and adjust the settings for a $[-2, 3] \times [-1, 12]$ window. Then press F5 to view the graph. The graph should have the same basic shape as the graph of $y = 4^x$.

Example: Graph $y = e^{-x/2}$ in the standard viewing window.

There are two things that you must remember when entering this function. Use the (-) key for the opposite of $\frac{x}{2}$ and enclose the exponent, $-\frac{x}{2}$, in parentheses. The shape of your graph should resemble the graph in Figure 22 reflected over the y-axis.

Turning Off the Axes

Sometimes it is helpful to view a graph without the presence of the x- and y-axes. To turn off the axes, press GRAPH MORE F3 for FORMT, highlight AxesOff, and then press ENTER. Try graphing one of the functions in the examples above with the axes turned off. (Remember to turn the axes back on when you are finished experimenting.)

SECTION 8: LOGARITHMIC FUNCTIONS

Graphing Logarithmic Functions

The logarithmic functions with base e and base 10 have their own function keys [LN] and [LOG], respectively. Logarithmic functions of other bases can be graphed by dividing these functions by the appropriate scaling factor.

Example: Graph $h(x) = \ln(x)$ in the window $[-1, 12] \times [-2, 3]$.

- Press [GRAPH] [F1] and [LN] [(] [F1] [)].
- Press [EXIT] [F2] and adjust the settings for a $[-1, 12] \times [-2, 3]$ window. Then press [F5] to view the graph.

SECTION 9: TRIGONOMETRIC FUNCTIONS

Graphing Trigonometric Functions

Three of the six basic trigonometric functions are built-in functions on the TI-85/86: sine [SIN], cosine [COS], and tangent [TAN]. Before graphing any of these functions, you should first check that your calculator is set in radian mode: press [2nd] [[MODE]]. If Radian is not already highlighted, move the cursor over Radian and press [ENTER].

Example: Graph $y = \sin(x)$ and $y = \csc(x)$ in the trigonometric viewing window.

- Press [GRAPH] [F1]. Enter the function $\sin(x)$ as $y1$ by pressing [SIN] [(] [F1] [)].
- Enter $\csc(x)$ as $y2$: Press [(] [SIN] [(] [F1] [)] [)] [2nd] [$[x^{-1}]$].

Note: *Since* $\csc(x)$ *is defined as* $1/\sin(x)$, *we use the reciprocal key,* [$[x^{-1}]$], *above.*

Warning! *The* [SIN^{-1}]*-key on your calculator is NOT the same as the csc function.*

- To graph these functions in the trig viewing window, press [EXIT] to access the GRAPH menu. Then press [F3] [MORE] [F3] for ZTRIG. Your graph should be similar to Figure 23.

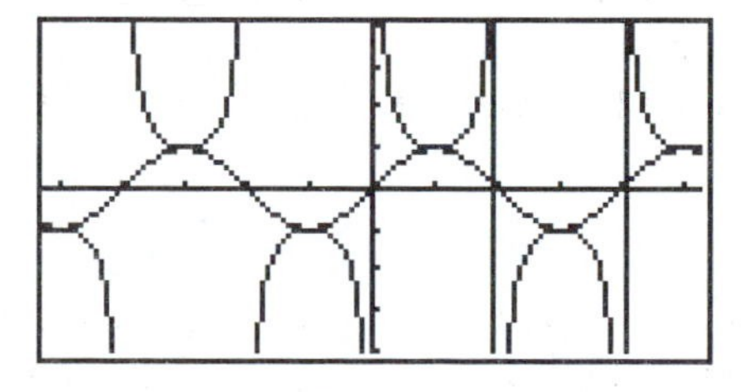

Figure 23 Graph of $\sin(x)$ and $\csc(x)$

Note: *The vertical lines indicate vertical asymptotes and are not part of the graph of* $y = \csc(x)$. *The TI-85/86 calculators only draw the vertical asymptotes that lie to the right of the y-axis. Where should the vertical asymptotes to the left of the y-axis be placed?*

- Press [F2] to observe the settings for the trig viewing window.

The trigonometric viewing window gives a good picture of $g(x)$. Keep in mind, however, that it is not the best window for viewing all trigonometric functions. For example, it would not be a good viewing window for the function $y = 5\cos(10x)$. For this function, you would need to adjust the window settings in order to display the key features of its graph.

Doing Away with Dull Trig Tables!

With the TI-85/86 you can solve problems in right-triangle trigonometry without using trig tables. To compute the sine, cosine, or tangent of an angle measured in degrees, first change the Radian/Degree mode setting to Degree: press [2nd] [[MODE]], highlight Degree, and press [ENTER]. Press [2nd] [[QUIT]] to return to the home screen.

Example: Compute sin 30°.

- Press [SIN] [3] [0] [ENTER]. (If you do not get 0.5 for the answer, go back and check that you have changed your calculator to degree mode.)

Example: Compute $\cos^{-1}(0.5)$ (Recall this is the angle whose cosine is 0.5.) and $\tan^{-1}(2.5)$.

- Press [2nd] [[COS⁻¹]] [.] [5] [ENTER]. Did you get 60? Remember this is 60° if you're in degree mode.
- Check that $\tan^{-1}(2.5) \approx 68.199°$.

Finally, here's a way to compute the sine, cosine, or tangent of an angle given in degrees without changing the mode setting. First, return your calculator to its default mode setting, Radian: press [2nd] [[MODE]], highlight Radian, and press [ENTER]. Press [2nd] [[QUIT]] to return to the home screen. Now, let's compute cos(60°) and sin(35°) without changing the mode setting. (This is a good idea because, for most of your work in precalculus, you want radian mode.)

- Press [COS] [6] [0] [2nd] [[MATH]] [F3], for angle, [F1], to insert the degree symbol, [ENTER]. You should get 0.5 for your answer.
- Now use this method for sin(35°). Did you get approximately 0.574?

SECTION 10: MULTIVARIABLE FUNCTIONS

Graphing a Function with More Than One Input Variable

You can represent functions with more than one input variable graphically by replacing one (or more) of the independent variables with a list of values. (If you don't remember how to handle lists, refer to *Graphing a Family of Functions* on page 266.)

Example: Examine the behavior of $F(w, x) = 3w - 2x$ by holding w constant, first at -2, then at 0, and then at 2, while x varies.

The three graphs, taken together, show how F varies with x for three different values of w. By forming a list $\{-2, 0, 2\}$ of the constant values for w, you can produce the graphs $y = F(-2, x)$, $y = F(0, x)$, and $y = F(2, x)$ using a single functional expression. Here's how:

- Press [GRAPH] [F1] and enter the function as $y1$ using the TI-85/86's list capabilities: press [3] [2nd] [[LIST]] [F1] for {, [(-)] [2] [,] [0] [,] [2] and [F2] for }. Then press [EXIT] (to return to the GRAPH/$y(x)$ = screen) followed by [–] [2] [F1]. Notice that the input variable w has been replaced by the list of values $\{-2, 0, 2\}$.
- Press [EXIT] [F3] [F4] and watch as three parallel lines are graphed one by one.

SECTION 11: PARAMETRIC EQUATIONS

Graphing Parametric Equations

For graphing parametric equations you'll need to change your calculator from function (Func) mode to parametric (Param) mode. Here's how: press [2nd] [[MODE]], then use the arrow keys to select Param and press [ENTER].

Now let's see the changes in the GRAPH menu and window screen.

- Press [GRAPH].
- Press [F1] to access the GRAPH/E(t) = menu. Notice that the input variable associated with the [F1] key is now t. (If you have any parametric equations stored in memory, erase them by positioning the cursor on each equation and pressing [CLEAR].)
- Press [EXIT] to return to the GRAPH menu. Press [MORE] [F3] for FORMT. Check that DrawLine is highlighted. Then press [EXIT].
- Press [F3] [F4] to set up the standard viewing window for parametric equations. Then press [F2] to observe the window settings. Use the up [↑] and down [↓] arrow keys to scroll through the entries in this menu. Notice that the settings associated with x and y are the same as they are in function mode. However, when your calculator is in parametric mode, you must also specify bounds and increments for the parameter t: tMin, tMax, and tStep.
- Press [2nd] [[QUIT]] to return to the home screen.

Example: Graph the set of parametric equations $x(t) = 2t + 1, y(t) = -3t + 5$ in the standard viewing window.

- Press [GRAPH] [F1] to enter the GRAPH/$E(t)$ = screen.
- Enter the equation for x opposite $xt1$: press [2] [F1] [+] [1] [ENTER].
- Enter the equation for y opposite $yt1$: press [(-)] [3] [F1] [+] [5] [ENTER].
- Press [EXIT] [F5] to graph the function. You should see a line segment in the right half of your screen.

- Press F4 for TRACE. The cursor will mark the location when $t = 0$. The value of t and the coordinates of the point will appear at the bottom of your screen. Press the right arrow key → and the cursor will jump to the location associated with $t \approx 0.13$ (t increases by one tStep). Press → repeatedly and watch the cursor move along the line.

You will use these equations again in the next example.

Example: Graph the position of a dot as it moves along the path $x(t) = 2t + 1$, $y(t) = -3t + 5$ at 0.5-second increments from time $t = 0$ seconds to $t = 3$ seconds.

For this example, we assume that your calculator is in parametric mode and that you have already entered this set of parametric equations in your calculator from the previous example.

- In the GRAPH menu, press MORE F3 for FORMT (format). Move the cursor to DrawDot and press ENTER.
- Press F2 and adjust the parameter settings for t: tMin = 0, tMax = 3, and tStep = .5.
- Press F4 for TRACE. Then press → and watch the dot move from one position to the next (on a screen similar to Figure 24) in 0.5 second time increments.

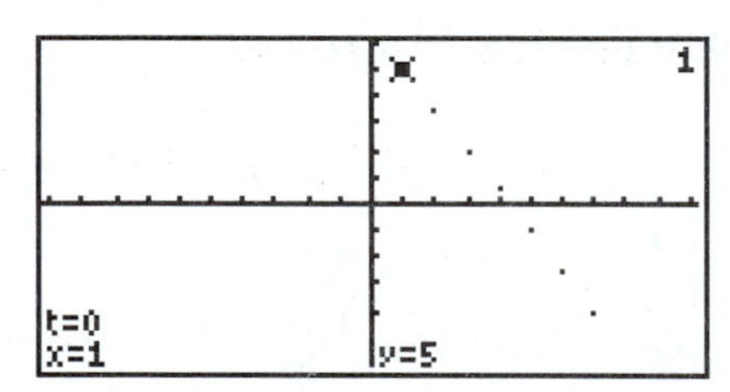

Figure 24 Graph in Dot Mode Setting

Combining Two Sets of Parametric Equations

In the lab for this section, you are asked to form a new set of parametric equations from two other sets of parametric equations.

Example: Suppose that you have two sets of parametric equations

$$S_1: \begin{array}{l} x_1 = 2t + 1 \\ y1 = -3t + 5 \end{array} \qquad S_2: \begin{array}{l} x_2 = t - 5 \\ y_2 = 4t - 3 \end{array}$$

and that you want to graph the combination $(1 - t)S_1 + tS_2$ over the interval $0 \leq t \leq 1$.

Step 1: Be sure that your calculator is in parametric mode. If you have not already done so, return your calculator to DrawLine format: press GRAPH MORE F3, select DrawLine, and press ENTER.

Step 2: Enter the two sets of parametric equations, S_1 and S_2, as *xt*1, *yt*1 and *xt*2, *yt*2.

Step 3: Enter the *x*- and *y*-equations for the combination $(1 - t)S_1 + tS_2$:

$$x_3 = (1 - t)x_1 + tx_2$$

$$y_3 = (1 - t)y_1 + ty_2$$

- The cursor should be opposite *xt*3. Enter the equation for *xt*3:

  Press [( ] [1] [−] [F1] [ )].

  Press [F2] for *xt* and then [1] to get *xt*1.

  Press [+] [F1].

  Press [F2] for *xt* and then [2] to get *xt*2.

  Press [ENTER].

- The cursor should be opposite *yt*3. Enter the equation for *yt*3:

  Press [( ] [1] [−] [F1] [ )].

  Press [F3] for *yt* and then [1] to get *yt*1.

  Press [+] [F1].

  Press [F3] for *yt* and then [2] to get *yt*2.

Step 4: Graph the combination $(1 - t)S_1 + tS_2$. Here's how:

- Unselect (turn off) parametric equations *xt*1, *yt*1, *xt*2, and *yt*2. Move the cursor opposite *xt*1. Press [F5]. Notice that both *xt*1 and *yt*1 have been unselected. (The highlighting over their equals signs has been removed.) Next, move the cursor opposite *xt*2 and press [F5].

- Press [EXIT] to return to the GRAPH menu and press [F2]. Set *t*Min = 0, *t*Max = 1, and *t*Step = 0.1. Adjust the remainder of the settings for a $[-5, 2] \times [-4, 6]$ window.

- Press [F5] for GRAPH. Your graph should resemble the one in Figure 25.

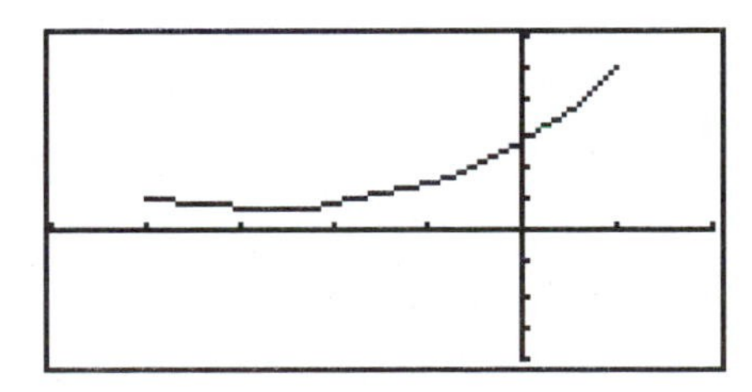

Figure 25 A Bézier Curve

Using Square Scaling in Parametric Mode

Here's how to get a window with square scaling in parametric mode.

- First, select a viewing window that shows the basic details of your graph.
- Then, in the GRAPH menu, change to square scaling by pressing [F3] [MORE] [F2] for ZSQR .

Example: Graph the set of parametric equations $x = 3\cos(t)$, $y = 3\sin(t) + 2$.

- If necessary, adjust the MODE and GRAPH/FORMT settings as follows. In the MODE menu (press [2nd] [MODE]), check that Radian and Param are highlighted. In the GRAPH/FORMT menu (press [GRAPH] [MORE] [F3]), check that SeqG and DrawDot are highlighted.
- Press [GRAPH] [F1] and erase any previously stored functions. Then enter the set of parametric functions above.
- Press [EXIT] to return to the GRAPH menu and then press [F2]. Set tMin to 0, tMax to 6.3, and tStep to 0.1. Adjust the remaining settings for a $[-6, 6] \times [-6, 6]$ window.
- Press [F5] to view the graph. Your graph should look egg-shaped. Next, press [F3] [MORE] [F2] for ZSQR and observe the graph in a window with square scaling. Your graph should look like a circle.

TI-92 GUIDE

This guide provides background on the TI-92 graphing calculator that will be useful for *Precalculus in Context: Projects for the Real World.* It consists of a basic tutorial followed by additional instructions relevant to each section in the laboratory manual.

BASIC TUTORIAL

Making Friends with the Keyboard

The layout of the keyboard is shown in Figure 1. It comprises:

- the function keys F1–F8 (to the left of the calculator screen)
- the Qwerty keyboard (below the calculator screen)
- calculator-type keys (on the right-hand side) including [ESC], [APPS], [ENTER], and the cursor keypad (8 directions: up, down, right, left, and diagonals)

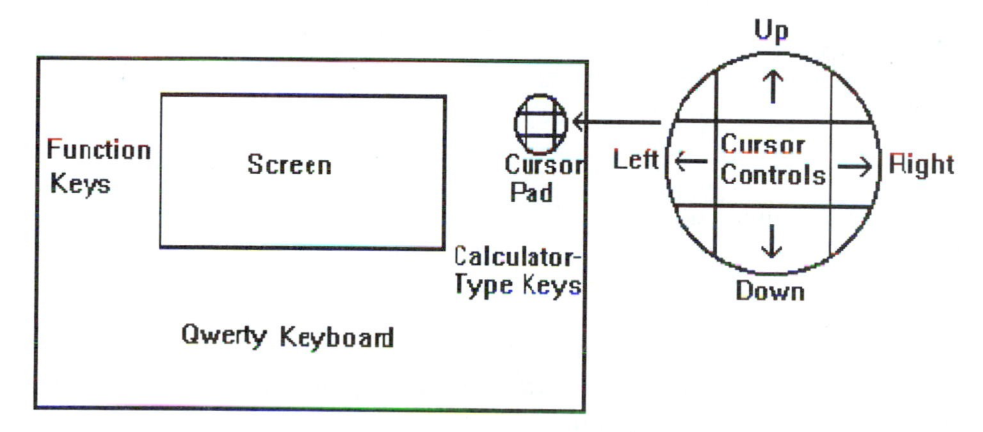

Figure 1 Layout of the Keyboard

The multiple function keys operate as follows:

- To access the function written in white lettering, press the key.
- To access the function written in green, first press the green diamond key.
  Example: [♦] [[Y=]] (same key as [W]).
- To access the function written in yellow, first press the yellow second key.
  Example: [2nd] [[QUIT]] (same key as [ESC]).
- To access caps, use the [↑] -key (above ON). (To lock and later to unlock caps, press [2nd] [[CAPS]].)

Getting Started: On, Off, and Contrast

Turn the calculator on by pressing [ON].

You may need to adjust the contrast. Hold down the [♦] and at the same time press [+] to darken or [−] to lighten.

To turn your calculator off, press [2nd][[OFF]]. If you forget, the calculator will automatically turn off after a period of non-use.

Calculating and Editing

You will be working mainly with the calculator-type keypad for this section of the tutorial. First, you need to understand something about the sections of your TI-92's screen. At the top of the screen in Figure 2 is a toolbar. Below the toolbar is the home screen, then the entry line (or command line), and at the very bottom the status.

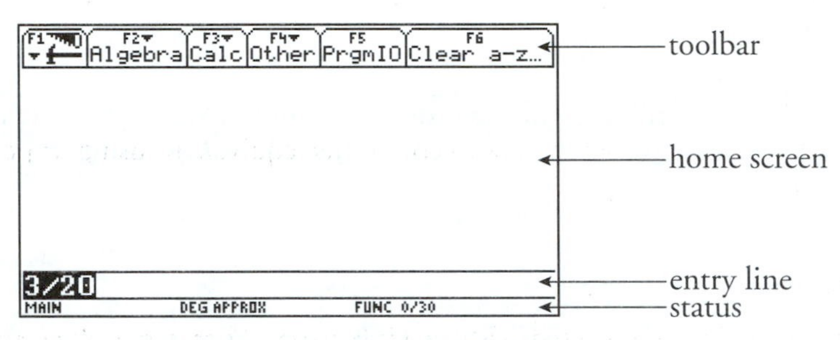

Figure 2 Section of TI-92 Screen

Press [F1] and then [8] to begin with a clear home screen. You do *not* have to clear the screen after each computation. If not cleared, the home screen will keep a history of your work.

Example: Compute 3×4.

From the calculator-type keypad, press [3][×][4] and then [ENTER]. (Any of the three [ENTER]-keys will work.) Note that the original problem, written as $3 * 4$, remains on the entry line. Directly above the entry line the original problem, written as $3 \cdot 4$, appears at the left side of the screen and the answer appears at the right. Press [CLEAR] to clear the entry line.

Warning! *The TI-92 has two minus keys, [−] and [(-)], to differentiate between the operation of subtraction (such as $3 - 2 = 1$) and the opposite of the positive number 2, namely -2. On your calculator screen, the subtraction sign appears slightly longer than the negative sign.*

Example: Compute $-2 + 5$.

To compute $-2 + 5$ press [(-)] (the key to the left of [ENTER] in the bottom row of the calculator-type keypad) and then [2], to create the number -2. Finish the computation to get the answer 3.

Note: *If you press [−] instead of [(-)] in this problem, you get the wrong answer! Try it for yourself.*

Example: Compute $5 - 2$.

To compute $5 - 2$, press [5][−][2][ENTER].

Example: Compute 8^2 and 1.05^7.

Press [8] [^] [2]. Now try 1.05^7 using [^] [7] to compute the power.

Example: Compute $\sqrt{16}$.

Press [2nd] [[√]] (same key as [×]) followed by [1] [6] [)] [ENTER]. Notice that a left parenthesis was automatically inserted after the square root sign. (If you forget to enter the right parenthesis, you'll get an error message.)

Example: Try to compute $\sqrt{-16}$.

If your calculator is set for real numbers, then you will get this error message: **Non-real result**. Press [ESC] to remove the error message from your screen. If the mode setting for Complex Format is set to Rectangular, then you would have gotten the answer $4i$. In that case, change the mode setting to Real and try this example again. (See *Changing Mode Settings* on page 286.)

Example: Compute $\sqrt[5]{32}$.

You will have to convert this problem to its equivalent using exponents: press [3] [2] [^] [(] [1] [÷] [5] [)] [ENTER].

Correcting an Error

We tackle two situations connected with making errors. First, we look at an example of an error that your calculator recognizes as an error. Then we provide an example that illustrates what you can do when you discover that you have punched in an error that the calculator is able to compute.

Correcting by Deleting

Example: Let's start by making a deliberate error: press [3] [+] [+] [2] [ENTER]. The message in Figure 3 will appear on your screen.

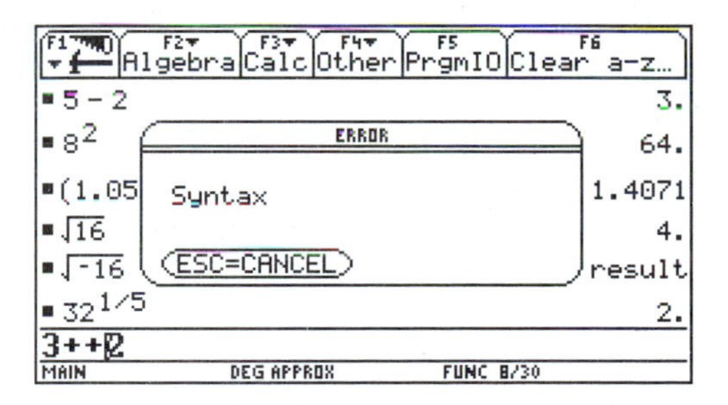

Figure 3 Error Message

Press [ESC] and the cursor will direct you to the error. Erase one of the plus signs by pressing [←] (this key, located at the bottom of the Qwerty keyboard, acts like a backspace key), and then press [ENTER]. The correct answer to 3 + 2 will appear.

Correcting by Inserting

Example: Press [3] [+] [4] [ENTER] and suppose that you really wanted $-33 + 4$. The expression $3 + 4$ on the entry line should be highlighted. Press the left-arrow on the cursor pad. This will move your cursor to the front of the entry $3 + 4$. Now press [(-)] [3] to insert -3 and then [ENTER] to compute the answer. You should get -29.

Resetting the Memory

Warning! *Resetting the memory erases all data and programs. All calculator settings will return to the default settings.*

Here's how to reset your calculator:

- Press [2nd] [[MEM]]. The screen should say MEMORY at the top.
- Press [F1] for reset. Your screen should match the one in Figure 4.

Figure 4 Reset Options

Memory clears the memory (programs, data, functions, graphs) but leaves the mode settings alone, **Default** leaves the memory alone but returns the mode settings to their defaults, and **All** resets both memory and defaults.

- If you have programs or data that you do not want to erase, press [3] to return only the mode settings to the defaults. Otherwise, press [1] to reset the memory and return all calculator settings to the defaults. The calculator will respond: **Reset ALL, are you sure?** or **Reset Defaults?** depending on whether you pressed 3 or 1. Press [ENTER] for YES. You will probably need to adjust the contrast by pressing [♦] and either [−] or [+].

Changing Mode Settings

Press [MODE]. The cursor should be blinking on FUNCTION. If you have reset your calculator as shown above, the mode settings will match the default settings in Figure 5. Notice that there are two pages to this menu. Press [F2] and then [F1] to move between pages 1 and 2.

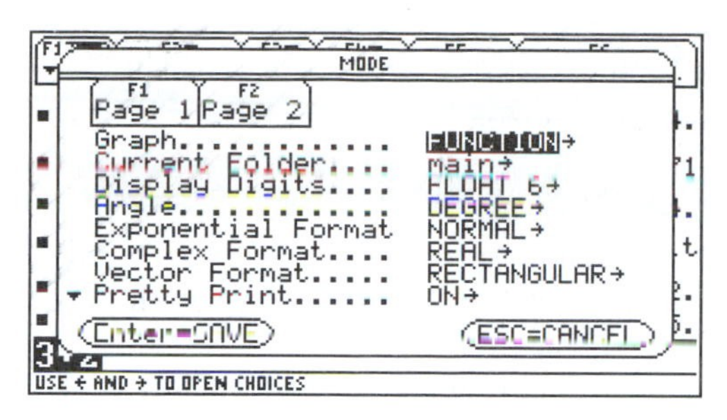

Figure 5 Mode Menu

Suppose that you want to work with complex numbers. Press the down-arrow on the cursor pad to highlight REAL (to the left of Complex Format). Next, press the right-arrow on the cursor pad to show the options. Press [2] to select RECTANGULAR followed by [ENTER] to save the setting.

Example: Compute $\sqrt{-16}$ while your calculator is set to RECTANGULAR.

The displayed answer should read: $4i$.

To return the mode setting to REAL, press [MODE]; press the down-arrow on the cursor pad to highlight RECTANGULAR, the right-arrow to show the options, and then press [1] [ENTER].

Graphing

Before getting started on a graphing example, check your mode settings: press [MODE]. For now, adjust the mode settings to match the default settings in Figure 5.

Next, check to see whether any functions have been stored in your calculator; press [APPS] [2] to view the Y= Editor. The screen in Figure 6, for example, indicates that two functions have been stored. Here's one way to erase these functions.

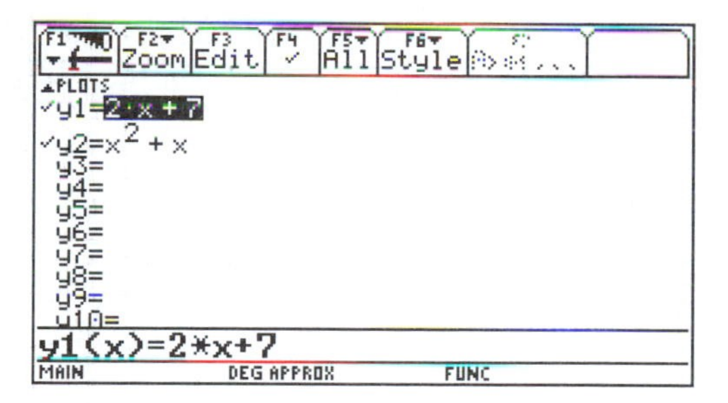

Figure 6 Y= Editor

- Use the cursor control keys to highlight a function. (In Figure 6, $y1$ is already highlighted.)
- Press [CLEAR] to erase the function.
- Press [2nd] [[QUIT]] to return to the home screen.

If you did not have any functions stored in your calculator, test these instructions after you have completed the next example. As you work through this guide, unless you are told otherwise, you should erase stored functions before you begin a new example.

Example: Graph $y = x^2$ and $y = x^3$ in the standard viewing window.

- Press [APPS] [2] (or [♦] [[Y=]]) for the Y= Editor. If the cursor (a black box) is not opposite $y1$, press the up-arrow and move it opposite $y1$.
- Enter the first function by pressing [X] [^] [2] followed by [ENTER]. (Note that the function will appear on the entry line until you press [ENTER].)
- The cursor should now be opposite $y2$. Press [X] [^] [3] [ENTER] to enter the second function.
- To graph $y = x^2$ and $y = x^3$ in the standard viewing window, press [F2] for the Zoom menu and [6] for ZoomStd. The graphs of both functions should appear in your screen.
- To check the window settings, press [APPS] [3] (or [♦] [[WINDOW]]). Your screen should match the one in Figure 7.

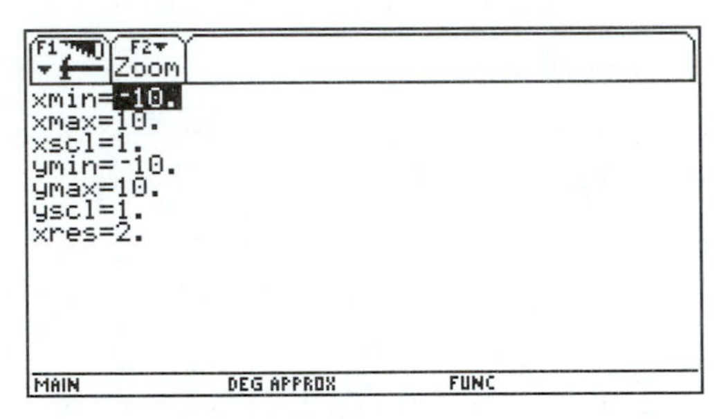

Figure 7 Window Editor

In the standard viewing window, the scaling on the x- and y-axes goes from -10 to 10 with tick marks one unit apart (since xscl and yscl = 1). We may refer to this window using the notation $[-10, 10] \times [-10, 10]$.

- To exit this menu and return to the home screen press [2nd] [[QUIT]] (same key as [ESC]).

In the next example, you will continue where the previous example left off.

Example: Graph $y = x^2$ and $y = x^3$ in the window $[-1.5, 1.5] \times [-1.0, 2.0]$. Place tick marks every 0.5 units. (In this example, the window settings should match the ones on the screen in Figure 8.)

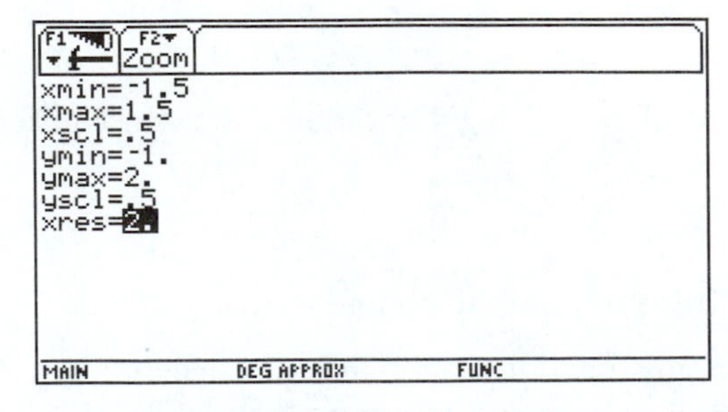

Figure 8 New Window Settings

- You should be looking at the window settings from the previous example. To change the setting for xmin, press [(-)] [1] [.] [5] [ENTER].
- Now, your cursor should be to the right of xmax. Change the value of xmax to 1.5 and press [ENTER]. Continue changing the window settings until yours match the ones in Figure 8.
- Finally, press [APP] [4] (or [♦] [[GRAPH]]) to view the graph in the new window.

When you have completed this example, erase the functions in $y1$ and $y2$. (See instructions at the beginning of *Graphing.*)

That's it! You have completed the tutorial. Now practice and experiment on your own with the calculator until you begin to feel comfortable with these basic operations. The remainder of this guide will introduce new techniques as they are needed, section by section, for your work in *Precalculus in Context: Projects for the Real World.*

Section-by-Section Guide

SECTION 1: LINEAR FUNCTIONS

Clearing Stored Functions

Unless you are told otherwise, erase any previously stored functions each time you begin a new example. The tutorial demonstrated one method for clearing functions. Here is another.

- Press [2nd] [[VAR-LINK]] (same key as [−]). You will see a list of all stored functions as well as lists, matrices, etc.
- Use the cursor pad to highlight the first function. Then press [F1] [1] for Delete and then [ENTER].
- Erase any remaining functions and then press [2nd] [[QUIT]] to return to the home screen.

Plotting Points

You can use your calculator to plot the Fahrenheit-Celsius data given in the preparation for Lab 1A. Then graph your guess for the formula that relates degrees Fahrenheit to degrees Celsius. This will allow you to check how closely the function specified by your formula follows the pattern of the data.

Example: Plot the data in the following table and then overlay the graph of $y = 18x + 85$.

Sample Data	
x	y
−2	40
−1	60
1	100
3	140

Step 1: Clear any stored functions. (See *Clearing Stored Functions.*)

Plan of action: On the TI-92, a data variable is a collection of lists. For example, the data variable in the table above would consist of an x-list and a y-list. You are allowed 99 lists. Each list will be placed in a separate column, C1–C99, of the Data/Matrix Editor. We'll store data from the x-column in C1 and the y-column in C2. The instructions for entering the data are in Step 2. The directions for plotting the data are outlined in Step 3.

Step 2: Enter the data. Here's how.

- Press [APPS] [6] for the Data/Matrix Editor.

- Press 3 for New. Your screen should match the one in Figure 9.

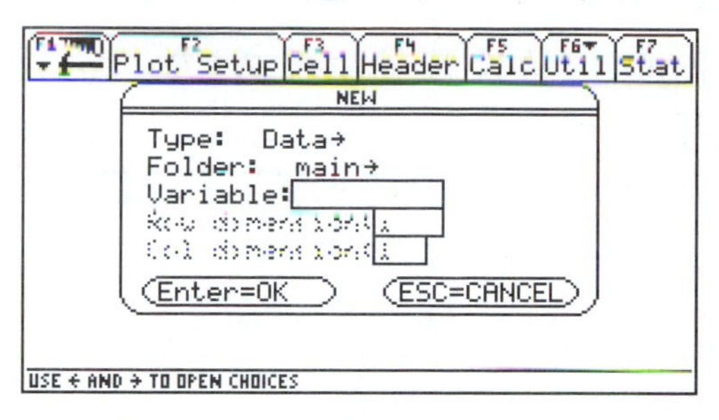

Figure 9 The Matrix/Data Editor

- Use the down arrow on the cursor pad to move the cursor to the box opposite Variable. Enter the name of your data set (the data variable): *sample*. Then press ENTER twice (once to enter the name and again to open a blank data table similar to the one in Figure 10).

Figure 10 Blank Data Table

- Press the up arrow on your cursor pad to highlight the cell *above* c1. Enter the name of the first variable, x, and press ENTER.
- Press the down arrow on the cursor pad to highlight the cell directly beneath c1 (as shown in Figure 10). Now enter the first column of numbers, pressing ENTER after each number.
- Next, use your cursor pad to highlight the cell above c2. Enter the name of the second variable, y, and press ENTER.
- Press the down-arrow on the cursor pad to highlight the cell directly below c2. Enter the y-data.

Step 3: Make a scatter plot of the data.

- Press F2 for Plot Setup.
- Plot 1 should be highlighted. Press F1 to define Plot 1. Your screen will match Figure 11. Enter c1 and c2 opposite x and y as indicated below and then press ENTER twice.

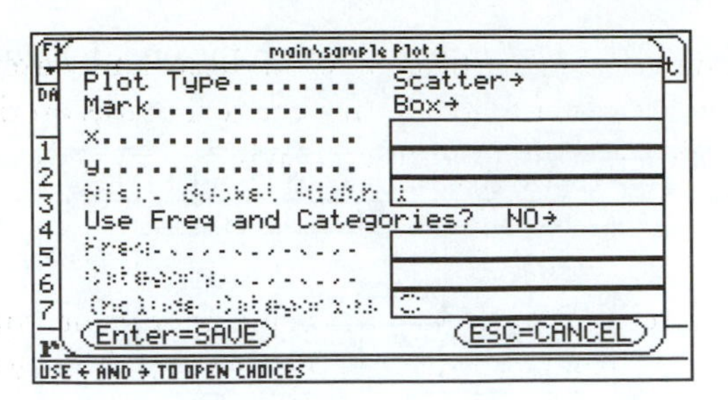

Figure 11 Defining Plot 1

- Press ◆ [Y=]. If necessary, press the up arrow on the cursor pad until you are able to see the Plot 1 entry (with a check to the left), above $y1 =$.
- Press F2 for Zoom and then 9 for ZoomData. When you use this option, your calculator selects a viewing window that displays all of your data.

Step 4: Enter the equation of the line, and graph the line and the scatter plot in the same viewing window.

- Press ◆ [Y=].
- Enter $18x + 85$ opposite $y1$.
- Press ◆ [GRAPH].

Warning! *If you fail to turn off Plot 1, this plot will appear superimposed on your next graph. Worse, if you subsequently erase the data variable without turning off this plot, you will get an error message* (**Undefined variable**) *each time you press* [GRAPH] *because your calculator will still be trying to create Plot 1.*

Step 5: Turn off the data plot. Press ◆ [Y=] F5 and select 5 (Data Plots Off). This will remove the check opposite Plot 1.

Step 6: Erase the data variable *sample.* Press 2nd [VAR-LINK], highlight sample, and then press F1 1 ENTER.

Adjusting the Viewing Window for Square Scaling

The viewing screen on your calculator is a rectangle. Therefore if you use the standard window, the tick marks on the y-axis will be closer together than those on the x-axis. For square scaling, we want the distance between 0 and 1 on the x-axis to be the same as the distance between 0 and 1 on the y-axis.

Example: Graph the line $y = x$ in the standard window and then switch to square scaling.

Step 1: Graph $y = x$ in the standard viewing window.

- Press ◆ [Y=] and enter $y = x$ as $y1$.
- Press F2 6 for ZStd.

Observe the spacing between the tick marks on the x- and y-axes. Notice that the tick marks on the y-axis are closer together than the tick marks on the x-axis and that the graph does not make a 45° angle with the x-axis.

Step 2: Change to a square viewing window.

- Press [F2] [5] to select ZSqr. When the graph appears on your screen, observe the equal distance between tick marks on the two axes. Notice that the graph makes a 45° angle with the x-axis.
- Press [♦] [[WINDOW]] and note the changes in the viewing window settings.

SECTION 2: QUADRATIC FUNCTIONS

Graphing a Quadratic Function

When you graph a quadratic function, it is important to experiment with various viewing windows, so as to ensure that you have captured all the important features of the graph on your screen.

Example: Graph $y = 2x^2 - 2x + 12$ using the three different viewing windows specified in the directions below.

Start by entering the function.

- Press [♦] [[Y=]] and enter $2x^2 - 2x + 12$. (Remember to use [−] when you subtract $2x$.)
- Now press [♦] [[WINDOW]].

Window 1: Adjust the settings to match those in Figure 12. Then press [♦] [[GRAPH]]. Your graph should be the familiar U-shape of a parabola.

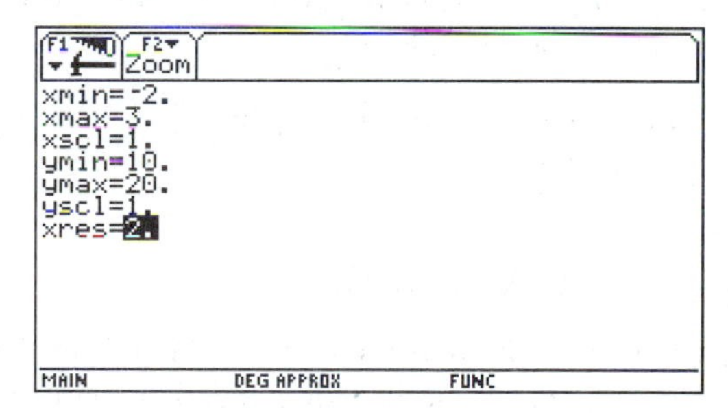

Figure 12 Settings for Window 1

Next, we view the graph of this quadratic function in two other windows. In each case, the viewing window selected fails to show key features of the parabola.

Window 2: Change the xmin setting to 2 as follows.

- Press [♦] [WINDOW]. The cursor should be opposite xmin. Change the value to 2.

- Press ♦ GRAPH to view the graph in this new window. In Window 2 the graph of $y = 2x^2 - 2x + 12$ looks more like a line than a parabola.

Window 3: Change to the standard viewing window by pressing F2 6. What do you see? In general, if you ask the calculator to graph a function, and just see empty axes, you probably are looking at a part of the plane that contains none of the graph.

Calculating Outputs of a Function

Example: Find the output of $f(x) = x^2 - x + 6$ when x has value 4.

- Press ♦ [Y=] and enter $x^2 - x + 6$ as $y1$.
- Press 2nd [QUIT] to return to the home screen.
- On the command line, enter $y1(4)$ and then press ENTER. The output should be 18.

Now, use the method outlined above to check that $f(-2) = 12$.

Using TRACE and ZoomBox to Approximate the Coordinates of a Point on the Graph

Example: Approximate the vertex (turning point) and x-intercepts of the parabola $y = 2x^2 + 3x - 2$.

Plan of action: First, we'll graph $y = 2x^2 + 3x - 2$ in the standard viewing window. Then we'll use ZBox to get a window that magnifies the section of the graph containing the vertex and x-intercepts. Finally, we'll use TRACE to estimate the vertex and x-intercepts.

Step 1: Enter and graph $y = 2x^2 + 3x - 2$ in the standard viewing window.

- Press ♦ [Y=] and enter the function $2x^2 + 3x - 2$ as $y1$.
- Press F2 6 to graph this function in the standard viewing window. Your graph should resemble the one in Figure 13 (without the box).

Step 2: Magnify the section of graph containing the vertex and x-intercepts by drawing a box that will become the new viewing window.

- Press F2 1 for ZoomBox. There should be a blinking cursor at the center of your screen.
- Next, we are going to draw a box like the one in Figure 13. Start by pressing the left and up-arrow keys on the cursor pad to move the cursor to the location where we want the upper left corner. Then press ENTER.
- Now, use the right and down-arrow keys to draw a box similar to the one in Figure 13. Press ENTER to replace the window settings with those that correspond to the box.

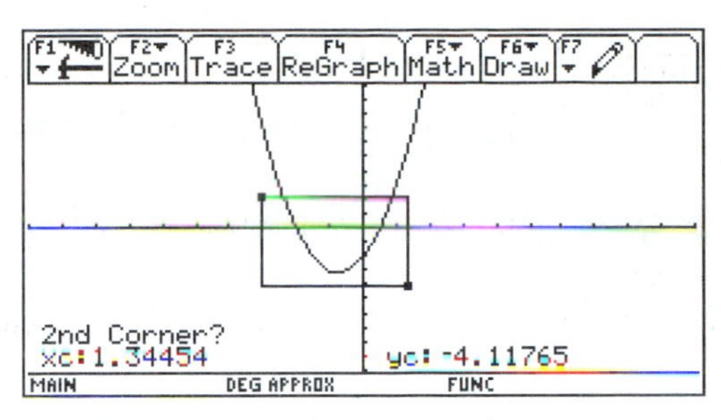

Figure 13 Graph of $y = 2x^2 + 3x - 2$

Step 3: Use TRACE to find the appropriate coordinates of the x-intercepts and vertex.

- Press [F3]. Use the right and left-arrow keys on your cursor pad to position the cursor at the vertex of the parabola. Then read the approximate coordinates of the vertex at the bottom of your screen.
- Continue to use the right [→] and left [←] arrow keys to position the cursor on each of the x-intercepts. Read the approximate coordinates for each of the x-intercepts.

Your answers should be similar to those in the screens in Figure 14.

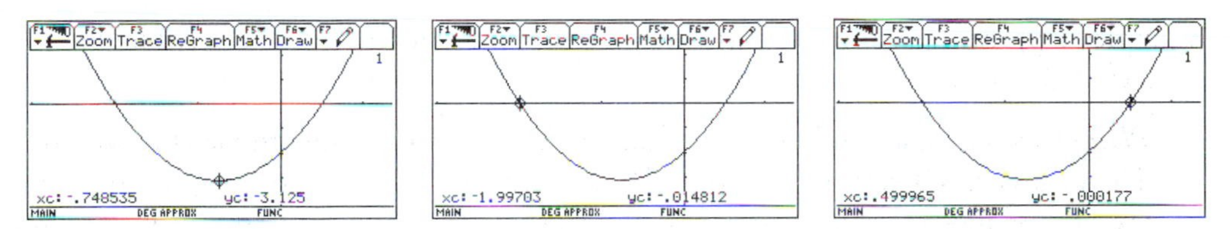

Figure 14 Approximate Coordinates of the Vertex and x-intercepts

SECTION 3: SHIFTING AND REFLECTING GRAPHS

In Lab 3, *Graph Trek Explorations*, you will be investigating the effect that certain algebraic modifications, such as adding a constant to the input variable, have on the graph of a function. You'll want to experiment using several different functions. We've provided some functions and algebraic modifications that you might want to consider.

Using Parentheses

Warning! *When you want to apply a function to an expression, you must enclose the entire expression in parentheses. For the built-in functions (such as the square root or absolute value function), the TI-92 will automatically insert the left parenthesis and you will need only to add the right parenthesis.*

Example: Graph $y = \sqrt{x + 2}$ in the window $[-5, 5] \times [-5, 5]$.

- In the Y= Editor, enter $y = \sqrt{x + 2}$ by pressing [2nd] [[√]] [X] [+] [2] [)]. (Note: the TI-92 automatically inserts the left parenthesis when you press the square root key.)

- Now, graph this function in the window $[-5, 5] \times [-5, 5]$. Your graph should resemble the one in Figure 15.

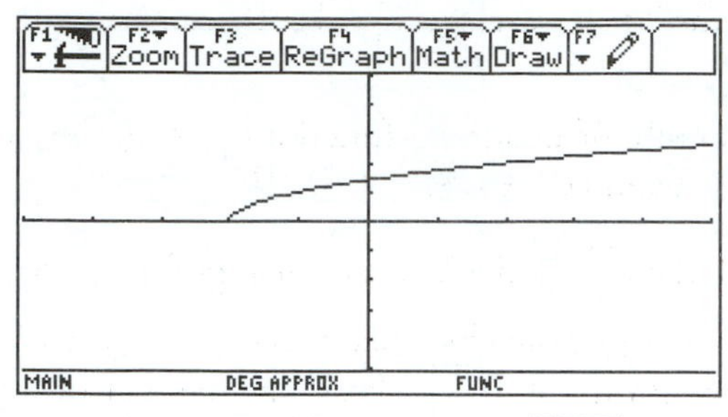

Figure 15 Graph of $y = \sqrt{x + 2}$

Graphing Trigonometric Functions in the Trig Viewing Window

Locate the SIN, COS, and TAN keys on your calculator. These keys will allow you to study the graphs of the sine, cosine, and tangent functions before the functions are formally introduced. Before entering any of these functions, press MODE. The Angle setting should be set to RADIAN. If it is, press 2nd [QUIT] to return to the home screen. If it is set to DEGREE, highlight that setting, press the right arrow on the cursor pad, press 1 for RADIAN, and then press ENTER.

Example: Examine the graphs of $y = \sin(x)$ and $y = \sin(x + 2)$ in the trig viewing window.

- Press ♦ [Y=] and enter the function $\sin(x)$ as $y1$: press SIN X) ENTER.
- Next, enter $\sin(x + 2)$ as $y2$.
- Press F2 7 for the trig viewing window. You should see two wavy curves. (If you don't, press MODE and check that the Angle setting is RADIAN.)
- Finally, press ♦ [WINDOW] and observe the settings for the trigonometric window.

Using the Absolute Value Function

Example: Graph $y = |x|$ in the standard viewing window.

- In the Y= Editor, enter $y = |x|$ as $y1$ as follows: Press 2nd [CATALOG]. The commands are listed in alphabetical order. If you are not already in the a's, press A. If necessary, use the up and down-arrow keys on your cursor pad to move the triangle marker opposite abs. Then press ENTER. Now complete entering the function by pressing X) ENTER.
- Press F2 6 to view the V-shaped graph of the absolute value function.

Graphing a Family of Functions

Using your calculator's list capabilities, you can substitute each value in a given list for a constant in an algebraic formula. This feature allows you to graph an entire family of functions quickly. On the TI-92, you specify a list by enclosing the members of the list in brackets: { }.

Example: Graph the family of quadratic functions $y = (x + 1)^2$, $y = (x + 2)^2$, and $y = (x + 3)^2$ in the window $[-5, 5] \times [-1, 10]$.

- Press [♦] [[WINDOW]] and adjust the settings for a $[-5, 5] \times [-1, 10]$ window.
- Enter the three functions by specifying the constants, 1, 2, and 3, in a list as follows. Press [♦] [Y=]. Position the cursor opposite $y1$ and press [(] [X] [+] [2nd] [[{]] [1] [,] [2] [,] [3] [2nd] [[}]] [)] [^] [2] [ENTER]. Then press [♦] [[GRAPH]] and watch as the three functions are graphed one after the other.

SECTION 4: EQUATIONS AND INEQUALITIES

Finding the Points of Intersection of Graphs

Example: Find the points where the graphs of $f(x) = -2x^2 + 2x + 12$ and $g(x) = -4x + 15$ intersect.

You could approximate the solutions to this problem using the techniques outlined in the example on page 294. However, the TI-92's intersect (ISECT) command will provide, when possible, exact solutions and, if not possible, very good approximations.

Step 1: Graph the functions in a viewing window that gives a clear view of the points of intersection.

- Enter the two functions into your calculator.
- Adjust the WINDOW settings so that you can see both points of intersection. (*Hint:* you might start with the standard viewing window, [F2] [6], and then adjust the WINDOW settings after looking at the graph.)

Step 2: Approximate the coordinates of one of the points of intersection.

- Press [F5] to access the Math menu and then press [5] for intersection. The image on your screen should be similar the one shown in Figure 16.

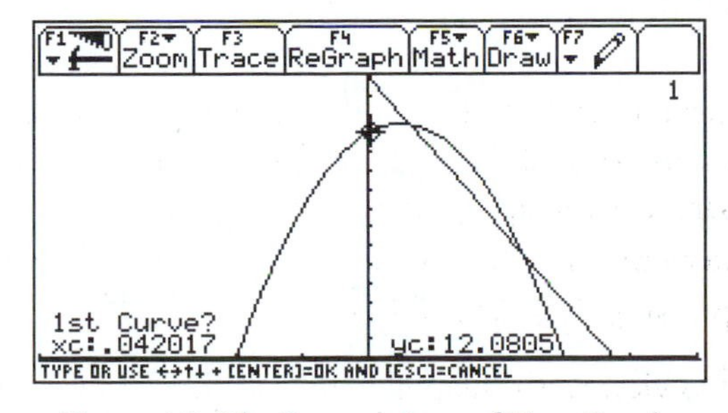

Figure 16 The Intersections of Two Graphs

- Try pressing the up- and down-arrow keys on the cursor pad. The cursor will jump back and forth from the line to the parabola. Position the cursor on the line (we'll designate the line to be the first curve) and press [ENTER]. The cursor should then jump to the parabola. Press [ENTER] to designate the parabola as the second curve.
- First, approximate the point of intersection having the smallest x-coordinate. You'll have to specify a narrow interval about this point of intersection. To specify the lower bound of this interval, use the left arrow on the cursor pad to position the cursor slightly to the left of this point of intersection and then press [ENTER]. To specify the upper bound of this interval, use the right arrow on the cursor pad to move the cursor slightly to the right of this point of intersection (but not so far that the second intersection point also lies in the interval). Then press [ENTER].

Step 3: Find the coordinates of the second point of intersection by repeating the entire process outlined in Step 2.

If you have done everything correctly, you will find that the two graphs intersect at approximately (0.634, 12.464) and (2.366, 5.536).

Graphing Piecewise Defined Functions

Example: Graph the piecewise defined function $f(x) = \begin{cases} x - 4 \text{ if } x > 4 \\ -x + 4 \text{ if } x \leq 4 \end{cases}$

The graph of $f(x)$ consists of two half-lines pieced together. You'll want the graph of $y = x - 4$ when x-values are greater than 4 and $y = -x + 4$ when x-values are less than or equal to 4.

Step 1: Press [♦][[Y=]] and enter each function along with its restriction as follows:

- The cursor should be opposite $y1$. Press [X][−][4] to enter the function. Next, enter the constraint on the domain: press [2nd][|] (the same key as K) followed by [X][2nd][>] (same key as [.]) [4]. Press [ENTER].
- The cursor should be opposite $y2$. Press [(-)][X][+][4] to enter the function. Next, enter the constraint on the domain: press [2nd][|] followed by [X][2nd][MATH][8] for test. Then press [4] to select ≤ followed by [4].
- Press [F2][6] to view the V-shaped graph of the two functions in the standard viewing window.

Step 3: Let's look at the graph of $y1$ alone by turning off the graph of $y2$.

- Return to the Y= Editor. Use the cursor pad to highlight the function opposite $y2$. Next, press [F4] to remove the check to the left of $y2$.
- Press [♦][[GRAPH]]. You should see only half of a line.
- Now turn $y2$ back on: Press [♦][[Y=]]. If necessary, use the cursor pad keys to highlight the function opposite $y2$. Press [F4] to restore the check mark to the left of $y2$.

Fitting a Line to Data

If your data, when plotted, lies exactly on a line, you can use algebra to determine the equation of the line. However, real data seldom fall precisely on a line. Instead, the plotted data may exhibit a roughly linear pattern. The least squares line (also called the regression line) is a line that statisticians frequently use when describing a linear trend in data.

Example: Use the least squares line to describe the linear pattern in the data below.

x	y
−3.0	−6.3
−2.0	−2.8
1.2	2.0
2.0	4.1
3.1	5.0
4.2	7.2

Step 1: Erase any stored functions and then enter the data.

- Erase any stored functions.
- Press [APP] [6] [3]. Enter a name for the data variable. Then press [ENTER] twice to bring up a blank data table. Enter your data, placing the data from the x-column in C1 and the y-column in C2. (For more specific details, refer to *Plotting Points* on pages 290–292.)

Step 2: Next, we'll find the equation for the least squares line. Your calculator will determine the values for the slope and intercept of $y = ax + b$.

Note: *Statisticians frequently specify the form of this line as* $\mathrm{y = ax + b}$, *where* a *and* b *represent the y-intercept and slope, respectively. However, your TI-92 reverses the roles of the constants* a *and* b *so that* b *is consistent with mathematicians' notation for linear functions,* $\mathrm{y = mx + b}$, *where* b *is the y-intercept.*

- Press [F5] to select the Calc menu. Next, press the right arrow on the cursor pad to show the choices for Calculation Type. Press [5] to select LinReg.
- Complete the remainder of the screen as follows:

Enter c1 for x and c2 for y. You'll want to store the regression equation: highlight the entry to the right of Store RegEQ To and press the left-arrow key on the cursor pad. Select $y1(x)$ for the storage location of the regression equation. When you are finished, your screen should match Figure 17.

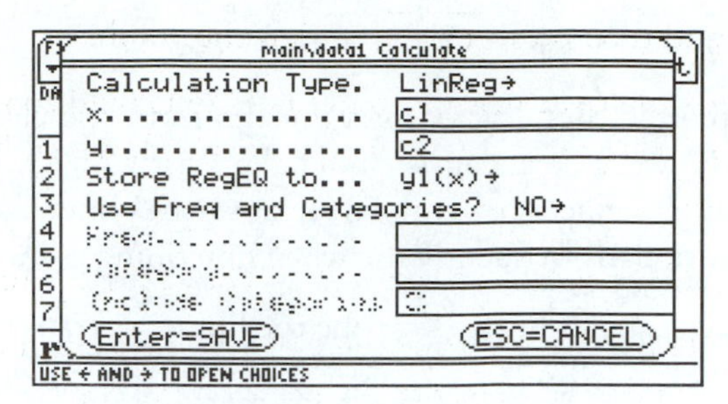

Figure 17 Calculation Screen

- Press ENTER. Your calculator will display the values for the slope and y-intercept of the regression (least squares) line. Press ENTER again.

Step 3: Make a scatter plot of the data and overlay a graph of the least squares line.

- Press F2 and highlight the plot number. Then press F1 and define the plot. (Refer to *Plotting Points* on pages 290–292.)
- After you have defined the plot, press ♦ Y=. There should be a check to the left of Plot 1 and the regression equation should be stored as $y1$. Now press F2 9 for ZoomData. Your screen should resemble Figure 18.

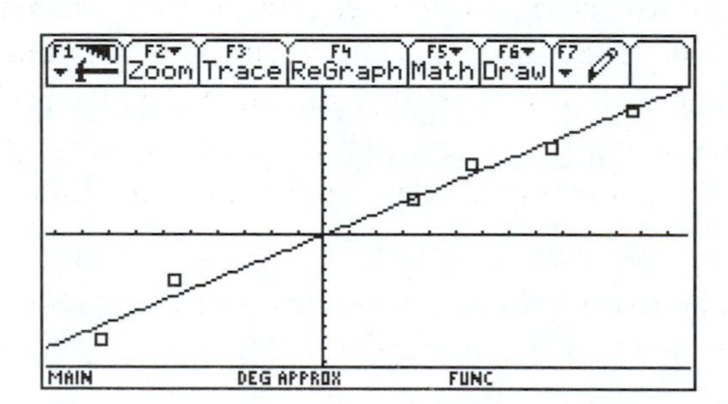

Figure 18 Scatter Plot and Regression Line

Warning! *Remember to turn off this plot. You can do this in the Y= Editor by pressing* F5 5 *for Data Plots Off.*

SECTION 5: POLYNOMIAL FUNCTIONS

Determining the Roots of a Function

Determining the roots (zeros, or x-intercepts) of a function using factoring often requires considerable skill and most polynomials can't be factored easily, if at all. However, your calculator can take much of the drudgery out of finding roots.

Example: Find the roots of the polynomial function $g(x) = \frac{1}{2}x^4 + 2x^3 + x^2 - 3x - 5$.

Step 1: Graph $g(x)$ in the standard viewing window.

Enter $g(x)$ as $y1$ and then press F2 6 to view the graph.

Step 2: Let's approximate the negative root first. (You could zoom in on this intercept and then use TRACE to estimate the x-coordinate. However, here is another way to determine the root.)

- Press F5 to bring down the Math menu. Then press 2 for zero.
- To approximate the negative root, you'll have to specify a narrow interval about the negative x-intercept. Press the left arrow on the cursor pad to move the cursor slightly to the left of the negative x-intercept and then press ENTER to mark with a black triangle the lower bound of the interval.
- Next, press the right arrow on the cursor pad to position the cursor slightly to the right of the negative root (*but still to the left of the positive root*) and press ENTER, this time marking the upper bound of the interval.
- Read off the approximate value of the root (or zero) from the bottom of your viewing screen. The x-value should be close to -3.05 and the corresponding y-value very close to zero. (Note that $-5E - 13$ is -0.0000000000005.)

Step 3: Adapt the instructions in Step 2 to approximate the positive root. After adapting the instructions, you should get approximately 1.39 for this root.

Finding Local Maxima or Minima of Functions

In Section 2, you found the vertex of a parabola using a procedure requiring a combination of ZoomBox and TRACE. (See page 294 for details.) The coordinates of any function's turning points could be estimated using this method. One drawback to this procedure is that you must frequently apply it several times in succession before you can obtain the desired accuracy. Here is another method for approximating local maxima and/or minima of a function.

Example: Let's estimate the local maximum and local minimum of the cubic function $f(x) = x^3 - 4x^2 + 2x - 4$.

Step 1: Graph $f(x)$ using a viewing window that gives you a clear view of the two turning points (one peak and one valley) of the graph.

Step 2: Approximate the coordinates of the turning point associated with the local maximum (the y-coordinate of the peak on the graph) as follows.

- Press F5 for Math menu and then 4 for Maximum.
- You'll have to specify a narrow interval about the turning point associated with the local maximum (the peak). Press the left arrow on the cursor pad to move the cursor slightly to the left of the peak and then press ENTER.
- Next, press the right arrow on the cursor pad to position the cursor slightly to the right of the turning point associated with the maximum and press ENTER.
- Read off the coordinates of this turning point at the bottom of your screen. The local maximum (y-coordinate of this turning point) should be approximately -3.73.

Step 3: Approximate the coordinates of the turning point associated with the local minimum (the y-coordinate of the valley on the graph).

- Press F5 3 for Minimum.
- Adapt the instructions for Step 2 to find the coordinates of the turning point associated with the local minimum (the valley). You should get a value for y that is close to -8.42.

SECTION 6: RATIONAL FUNCTIONS

Graphing a Rational Function

Warning! *If the numerator or the denominator of a rational function consists of more than one term, you must enclose it in parentheses when you enter it into your calculator.*

Example: Graph $r(x) = \dfrac{x^2 - 1}{x - 3}$.

- Press ♦ [Y=] and enter $r(x)$ as $y1$: press (X ^ 2 − 1) ÷ (X − 3) and then ENTER. If your calculator is in Pretty Print Mode, the function for $y1$ should appear as it is written in the example above.
- Press F2 6 to view the graph in the standard window. Your graph should look similar to the one in Figure 19.

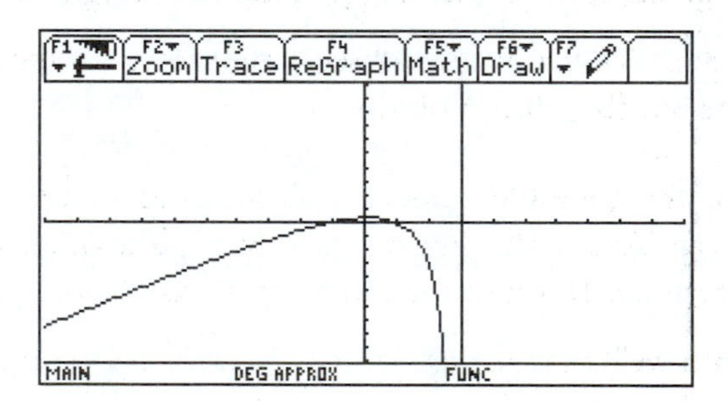

Figure 19 Graph of $r(x)$ in Standard Viewing Window

The vertical line in the graph above indicates that this function has a vertical asymptote at $x = 3$. Remember, this line is not part of the graph of the function. Furthermore, because the domain of this function is all real numbers except $x = 3$, there is a branch of this graph that lies to the right of the line $x = 3$. In order to observe this branch, you would have to adjust the window setting for $y$max.

Do *not* clear this function from your calculator's memory until after you have completed the next topic, *Zooming Out.*

Zooming Out

Example: Graph the function $f(x) = \dfrac{x^2 - 1}{x - 3}$ in the standard viewing window and then zoom out by a factor of 4 several times.

For this example, observe what happens to the appearance of the graph of $r(x)$ as we "back away" by increasing the width and height of the viewing window. You should have the graph of $r(x)$ (from the previous example) in your screen.

- Press [♦] [[WINDOW]] to access the window settings. Set xScl and yScl equal to 0. (This turns off the tick marks that appear on the axes. If you skip this step, the axis will get crowded with tick marks when you zoom out.) Press [♦] [[GRAPH]] to return to the graph.
- Check the settings for the zoom factor: press [F2] [C] for Set Factors. If necessary, adjust the factor settings so that xFact = 4 and yFact = 4 and then press [ENTER].
- Press [F2] [3] for ZoomOut. (A blinking pixel, part of a free moving cursor, should appear in the center of your screen. The zooming will be centered around this location. If you wish to change the focal point of the zooming, use the arrow keys to position this cursor at a different center.) Press [ENTER] to view the graph over wider x- and y-intervals. Wait until the cursor reappears in the center of your screen.
- Repeat the previous step to zoom out a second time. (The graph should look like a line except, perhaps, for a small blip slightly to the right of the origin.)
- Now, press [♦] [[WINDOW]] to observe the effect on the window settings of twice zooming out (by a factor of 4).

The default setting for ZoomOut widens both the x- and y-intervals by a factor of 4 each time that it is applied. In the previous example, you zoomed out twice. Therefore, the x- and y-intervals are 16 times wider than they were before you zoomed out.

In the next example, we'll change the default zoom settings in order to observe a graph of a function that begins to act like its horizontal asymptote.

Example: Graph $q(x) = \dfrac{5x^2 + 20x - 105}{2x^2 + 2x - 60}$ in the standard viewing window.

If you have entered the function correctly, your graph should look like Figure 20.

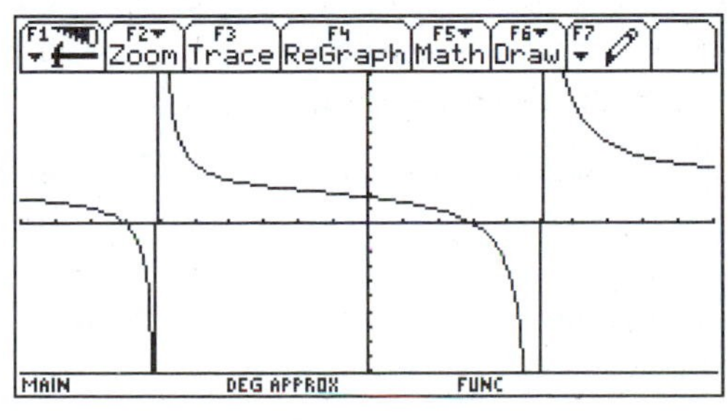

Figure 20 Graph of $q(x)$

Based on this graph, you may suspect that the function $q(x)$ has a horizontal asymptote but this is not at all obvious. Let's observe the function's graph over increasingly wide x-intervals to see if it begins to behave like a horizontal line. Instructions on how to zoom out in the horizontal direction follow.

- Press ♦ [WINDOW] and set xScl and yScl to 0. Now, press F2 C. Leave xFact set at 4; change the setting for yFact to 1. Then press ENTER twice.
- Press F2 3 ENTER to widen the x-interval by a factor of four. Repeat this process several more times to continue widening the x-interval. Your graph should begin to resemble its horizontal asymptote $y = 2.5$.

SECTION 7: EXPONENTIAL GROWTH AND DECAY

Graphing Exponential Functions

Because exponential functions increase or decrease very quickly in certain regions of their domain, you may have to experiment in order to find a viewing window that captures the function's important graphical features.

Example: Graph $y = 4^x$ in the standard viewing window. Then change the window to $[-2, 2] \times [0, 10]$.

- Press ♦ [Y=]. Enter $y = 4^x$ as $y1$ by pressing 4 ^ X.
- Press F2 6 to graph this function in the standard viewing window.

Notice that, to the left of the y-axes, the graph appears to merge with the line $y = 0$ and, to the right of the y-axes, the graph becomes so steep that it appears nearly vertical.

- Press ♦ [WINDOW] to access the window settings. Now adjust the settings for a $[-2, 2] \times [0, 10]$ window so that the graph will fill more of the screen. Your graph should resemble the one in Figure 21.

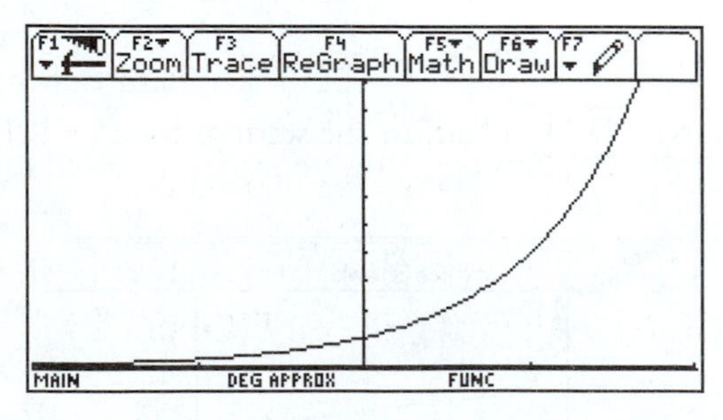

Figure 21 Graph of $y = 4^x$

There is one base for exponential functions that is so common that it has its own function key on the TI-92 calculator: $[e^x]$.

Example: Graph $f(x) = e^x$ in the window $[-2, 3] \times [-1, 12]$.

- Press ♦ [Y=] and enter $f(x)$: press 2nd [e^x] (same key as LN) X) ENTER .
- Press ♦ [WINDOW] and adjust the settings for a $[-2, 3] \times [-1, 12]$ window. Then press ♦ [GRAPH] to view the graph. The graph should have the same basic shape as the graph of $y = 4^x$.

Example: Graph $y = e^{-\frac{x}{2}}$ in the standard viewing window.

There are two things that you must remember when entering this function. Use the (-) key for the opposite of $\frac{x}{2}$, and enclose the exponent, $-\frac{x}{2}$, in parentheses. In shape, your graph should resemble the graph in Figure 21 reflected over the y-axis.

Turning Off the Axes

Sometimes it is helpful to view a graph without the presence of the x- and y-axes. To turn off the axes, you must be in the Y= Editor, Window Editor, or Graph screen. Press F1 9 for Format, change the setting for Axes from ON to OFF, and then press ENTER . Try graphing one of the functions in the examples above with the axes turned off. (Remember to turn the axes back on when you are finished experimenting.)

SECTION 8: LOGARITHMIC FUNCTIONS

Graphing Logarithmic Functions

The natural logarithmic function (base e) has its own function key LN . The common logarithmic function (base 10) can be accessed from the CATALOG or keyed-in directly using the Qwerty keypad. Logarithmic functions of other bases can be graphed by dividing these functions by the appropriate scaling factor.

Example: Graph $h(x) = \ln(x)$ and $g(x) = \log(x)$ in the window $[-1, 12] \times [-2, 3]$.

- Press ♦ [Y=] .
- To enter $h(x)$ as $y1$: press LN X) ENTER .
- To enter $g(x)$ as $y2$: press L O G (X) ENTER .
- Press ♦ [WINDOW] and adjust the settings for a $[-1, 12] \times [-2, 3]$ window. Then press ♦ [GRAPH] to view the two graphs.

SECTION 9: TRIGONOMETRIC FUNCTIONS

Graphing Trigonometric Functions

Three of the six basic trigonometric functions are built-in functions on the TI-92: sine SIN , cosine COS , and tangent TAN . Before graphing any of these functions, you should first check that your calculator is set in radian mode: press MODE . The Angle setting should be set to RADIAN. (If it is set to DEGREE, highlight that setting, press the right arrow on the cursor pad, press 1 for RADIAN, and then ENTER .)

Example: Graph $y = \sin(x)$ and $y = \csc(x)$ in the trigonometric viewing window.

- Press ♦ Y= and erase any stored functions. Then enter the function $\sin(x)$ as $y1$ by pressing SIN X).

  Note: *Since* $\csc(x)$ *is defined as* $\frac{1}{\sin(x)}$, *we will use the reciprocal key,* $[x^{-1}]$, *to enter* $\csc(x)$.

  Warning! *The* SIN^{-1}*-key on your calculator is NOT the same as the csc function.*

- Enter $\csc(x)$ as $y2$: press (SIN X)) 2nd $[x^{-1}]$ (the same key as 9) and then ENTER.
- To graph these functions in the trig viewing window, press F2 7 for ZoomTrig. Your graph should be similar to Figure 22.

Note: *The vertical lines indicate vertical asymptotes and are not part of the graph of* $y = \csc(x)$.

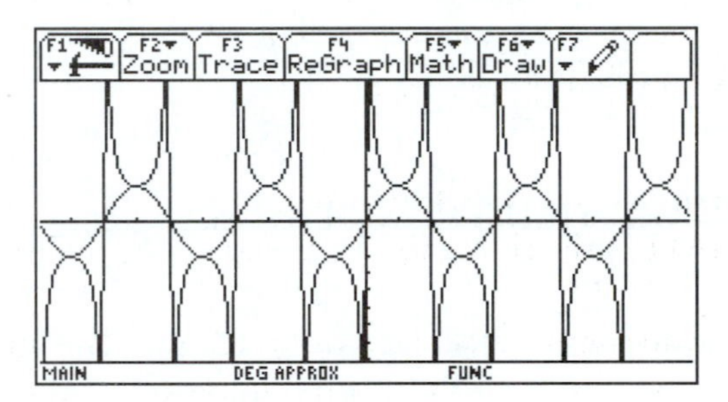

Figure 22 Graph of $\sin(x)$ and $\csc(x)$

- Press ♦ [WINDOW] to observe the settings for the trig viewing window.

The trigonometric viewing window gives a good picture of $g(x)$. Keep in mind, however, that it is not the best window for viewing all trigonometric functions. For example, it would not be a good viewing window for the function $y = 5\cos(10x)$. For this function, you would need to adjust the window settings in order to display the key features of its graph.

Doing Away with Dull Trig Tables!

With the TI-92 you can solve problems in right-triangle trigonometry without using trig tables. To compute the sine, cosine, or tangent of an angle measured in degrees, first change the Angle mode setting to DEGREE: press MODE. Then press the down arrow on the cursor pad to highlight RADIAN and then the right arrow on the cursor pad to show the options. Press 2 to select DEGREE and then ENTER to save the setting. If needed, press 2nd [QUIT] to return to the home screen.

Example: Compute $\sin 30°$.

- Press SIN 3 0 ENTER.

- Oops! If you fail to enclose the 30 in parentheses, you will get an error message. Press ESC. Now press) ENTER. (If you do not get $\frac{1}{2}$ or 0.5 for the answer, go back and check that you have changed your calculator to degree mode.)

Example: Approximate tan 25°.

- Press TAN 2 5) ENTER. Don't be surprised if you get tan(25). That's the exact answer. To get the decimal approximation, you'll need to change the Exact/Approx mode setting to APPROXIMATE.
- Press MODE F2, highlight the setting to the right of Exact/Approx, press the right arrow on the cursor pad to display the choices, and then press 3 for APPROXIMATE. Press ENTER to save the setting.
- tan(25) should still be highlighted on the command line. Press ENTER. Did you get approximately 0.466?

Example: Compute $\cos^{-1}(0.5)$ (recall this is the angle whose cosine is 0.5) and $\tan^{-1}(2.5)$.

- Press 2nd [COS] .) ENTER. Did you get 60? Remember this is 60° if you're in degree mode.
- Check that $\tan^{-1}(2.5) \approx 68.199°$. (If you don't get this answer check to see if APPROXIMATE is the choice for the Exact/Approx mode setting.)

Finally, here's a way to compute the sine, cosine, or tangent of an angle given in degrees without changing the mode setting. First, return your calculator to its default mode setting, RADIAN: press MODE, highlight DEGREE, press the right arrow on the cursor pad, and press 1 and then ENTER to select RADIAN. Now, let's compute cos(60°) and sin(35°) without changing the mode setting. (This is a good idea because, for most of your work in precalculus, you want RADIAN mode.)

- Press COS 6 0 2nd [MATH] 2, for angle, 1, to insert the degree symbol,) ENTER. You should get 0.5 for your answer.
- Now use this method for sin(35°). Did you get approximately 0.574?

You may wish to return the Exact/Approx mode setting to AUTO. Press MODE F2, highlight APPROXIMATE, press the right-arrow on the cursor pad to show the choices, select 1, and press ENTER.

SECTION 10: MULTIVARIABLE FUNCTIONS

Graphing a Function with More Than One Input Variable

You can represent functions with more than one input variable graphically by replacing one (or more) of the independent variables with a list of values. (If you don't remember how to handle lists, refer to *Graphing a Family of Functions* on page 297.)

Example: Examine the behavior of $F(w, x) = 3w - 2x$ by holding w constant, first at -2, then at 0, and then at 2, while x varies.

The three graphs, taken together, show how F varies with x for three different values of w. By forming a list $\{-2, 0, 2\}$ of the constant values for w, you can produce the graphs $y = F(-2, x)$, $y = F(0, x)$, and $y = F(2, x)$ using a single functional expression. Here's how:

- Press [♦] [[Y=]]. Enter the function using the TI-92's list capabilities: press [3] [2nd] [{] [(-)] [2] [,] [0] [,] [2] [2nd] [}] [−] [2] [X]. Notice that the input variable w has been replaced by the list of values $\{-2, 0, 2\}$.
- Press [ENTER] [F2] [6] and watch as three parallel lines are graphed one by one.

SECTION 11: PARAMETRIC EQUATIONS

Graphing Parametric Equations

For graphing parametric equations, you'll need to change your calculator from function mode to parametric mode. Here's how. Press [MODE]. The cursor should be blinking on FUNCTION (to the right of Graph). Press the right arrow on the cursor pad to display the choices and then press [2] to select PARAMETRIC. Press [ENTER] to save the setting. Now let's see the changes in the Y= Editor and WINDOW screen.

- Press [♦] [[Y=]]. Notice that the parametric equations come in pairs, for example: $xt1$, $yt1$. (If you have any parametric equations stored in memory, erase them by positioning the cursor on each equation and pressing [CLEAR].)
- Press [F2] [6] to set up the standard viewing window for parametric equations. Then press [♦] [[WINDOW]] to observe the window settings. Notice that the settings associated with x and y are the same as they are in function mode. However, when your calculator is in parametric mode you must also specify bounds and increments for the parameter t: $t$min, $t$max, and $t$step.
- Press [2nd] [[QUIT]] to return to the home screen.

Example: Graph the set of parametric equations $x(t) = 2t + 1$, $y(t) = -3t + 5$ in the standard viewing window.

- Press [♦] [[Y=]]. If necessary, move the cursor opposite $xt1$.
- Enter the equation for x: press [2] [T] [+] [1] [ENTER].
- The cursor should be opposite $yt2$. Enter the equation for y: press [(-)] [3] [T] [+] [5] [ENTER].
- Press [♦] [[GRAPH]] to graph the function. You should see a line segment in the right half of your screen. (If you see dots rather than a solid line, press [♦] [[Y=]], highlight the equation for $xt1$, press [F6] for Style and then [1] to select Line.)

- Press F3 for TRACE. The cursor will mark the location when $t = 0$. The value of t and the coordinates of the point will appear at the bottom of your screen. Press the right arrow on the cursor pad and the cursor will jump to the location associated with $t \approx 0.13$ (t increases by one $t$step). Press the right arrow on the cursor pad repeatedly and watch the cursor move along the line.

You will use this set of equations in the next example.

Example: Graph the position of a dot as it moves along the path $x(t) = 2t + 1$, $y(t) = -3t + 5$ at 0.5-second increments from time $t = 0$ seconds to $t = 3$ seconds.

For this example, we assume that your calculator is in parametric mode and that you have already entered this set of parametric equations in your calculator from the previous example.

- In the Y= Editor, highlight the equation for $xt1$. Then press F6 for Style and choose 2 for Dot.
- Press ♦ [WINDOW] to access the window settings. Adjust the parameter settings for t: $t$min = 0, $t$max = 3, and $t$step = 0.5.
- Press ♦ [GRAPH]. You should see a group of isolated dots that fall on a line. Press F3 for Trace. Then press the right-arrow on the cursor pad and watch the dot move from one position to the next (on a screen similar to Figure 23) in 0.5-second time increments.

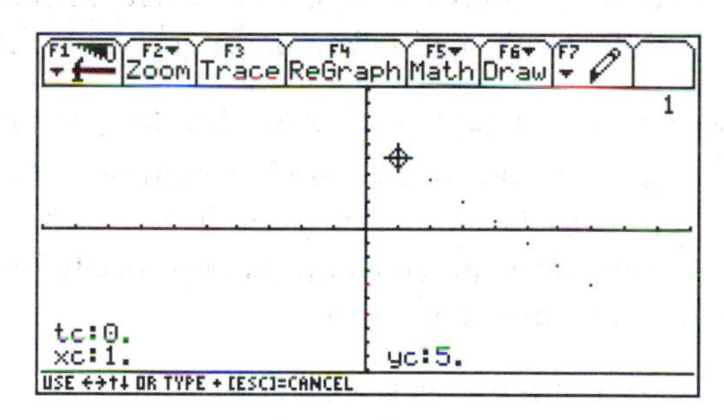

Figure 23 Graph in Dot Style

Combining Two Sets of Parametric Equations

In the lab for this section, you are asked to form a new set of parametric equations from a combination of two other sets of parametric equations.

Example: Suppose that you have two sets of parametric equations

$$S_1: \quad x_1 = 2t + 1 \qquad S_2: \quad x_2 = t - 5$$
$$\qquad y_1 = -3t + 5 \qquad \qquad y_2 = 4t - 3$$

and that you want to graph the combination $(1 - t)S_1 + tS_2$ over the interval $0 \leq t \leq 1$.

Step 1: Be sure that your calculator is in parametric mode. If you have not already done so, return your calculator to line style: press [F6] [1] in the Y= Editor.

Step 2: Enter the two sets of parametric equations, S_1 and S_2, as *xt*1, *yt*1 and *xt*2, *yt*2. (If you worked the previous example, the first set of equations already should be entered.)

Step 3: Enter the *x*- and *y*- equations for the combination $(1 - t)S_1 + tS_2$:

$$x_3 = (1 - t)x_1 + tx_2$$

$$y_3 = (1 - t)y_1 + ty_2$$

- The cursor should be opposite *xt*3. Enter the equation for *xt*3:
  - Press [(] [1] [−] [T] [)] [×].
  - Press [X] [T] [1] [(] [T] [)] for $x1(t)$.
  - Press [+] [T] [×].
  - Press [X] [T] [2] [(] [T] [)] for $x2(t)$.
  - Press [ENTER].
- The cursor should be opposite *y*3*t*. Enter the equation for *yt*3:
  - Press [(] [1] [−] [T] [)] [×].
  - Press [Y] [T] [1] [(] [T] [)] for $y1(t)$.
  - Press [+] [T] [×].
  - Press [Y] [T] [2] [(] [T] [)] for $y2(t)$.
  - Press [ENTER].

Step 4: Graph the combination $(1 - t)S_1 + tS_2$. Here's how:

- Unselect (turn off) parametric equations *xt*1, *yt*1, *xt*2, and *yt*2. Move the cursor opposite *xt*1. Press [F4]. Repeat this process to unselect *yt*1, *xt*2, and *yt*2. (Notice that the checks to the left of these equations have disappeared.)
- Press [♦] [[WINDOW]]. Set *t*min = 0, *t*max = 1, and *t*step = 0.1. Adjust the remainder of the settings for a $[-5, 2] \times [-4, 6]$ window.
- Press [♦] [[GRAPH]] to graph the combination. Your graph should resemble the one in Figure 24.

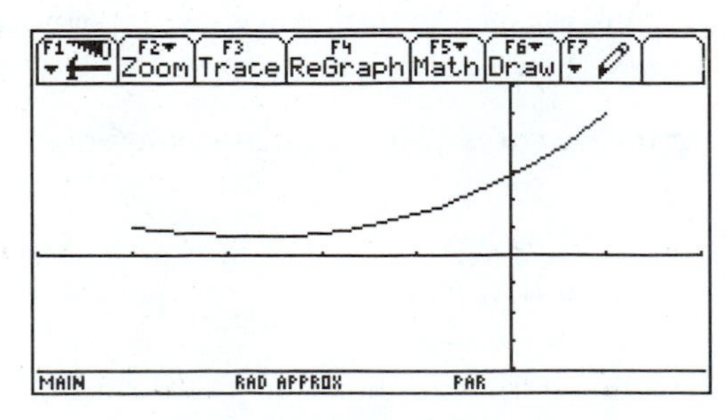

Figure 24 A Bézier Curve

Using Square Scaling in Parametric Mode

Here's how to get a window with square scaling in parametric mode.

- First, select a viewing window that shows the basic details of your graph.
- Then, in the Y= Editor, change to square scaling by pressing [F2] [5] for ZoomSqr.

Example: Graph the set of parametric equations $x = 3\cos(t)$, $y = 3\sin(t) + 2$.

- If necessary, adjust the MODE: select PARAMETRIC for the Graph setting and RADIAN for the Angle setting.
- Press [♦] [[Y=]] and erase any previously stored equations. With the cursor opposite $xt1$, press [F6] [1].
- Next, enter the set of parametric functions above.
- Press [♦] [[WINDOW]]. Set $t$min to 0, $t$max to 6.3, and $t$step to 0.1. Adjust the remaining settings for a $[-6, 6] \times [-6, 6]$ window.
- Press [♦] [[GRAPH]] to view the graph. Your graph should look egg-shaped. Next, press [F2] [5] for ZoomSqr and observe the graph in a window with square scaling. Your graph should look like a circle.

DATE DUE